Environmental Science and Engineering

Environmental Engineering

For further volumes:
http://www.springer.com/series/7487

Fokion K. Vosniakos

Radioactivity Transfer in Environment and Food

Springer

Fokion K. Vosniakos
Department of Sciences
Alexander Technological Educational Institute of Thessaloniki
57400 Thessaloniki
Greece

ISSN 1863-5520
ISBN 978-3-642-28740-4 ISBN 978-3-642-28741-1 (eBook)
DOI 10.1007/978-3-642-28741-1
Springer Heidelberg New York Dordrecht London

Library of Congress Control Number: 2012934576

Printed on acid-free paper

Springer is part of Springer Science+Business Media (www.springer.com)

Foreword

Our world is radioactive and has been since it was created. Over 60 radionuclides (radioactive elements) can be found in nature, and they can be placed in three general categories:

1. Primordial—from before the creation of the Earth
2. Cosmogenic—formed as a result of cosmic ray interactions
3. Human produced-enhanced or formed due to human actions (minor amounts compared to natural)

Radioactive substances are those that are unstable in nature. Radioactive types of uranium, thorium and radium emit radiation to reach a more stable condition. This process is called radioactive decay. By measuring the type of radiation emitted, the specific energy level or levels of radiation and the precise rate of decay, scientists are able to identify a radioactive substance and determine how much of it is present.

Radionuclides are found naturally in air, water and soil. They are even found in us, being that we are products of our environment. Every day, we ingest and inhale radionuclides in our air and food and the water. Natural radioactivity is common in the rocks and soil that makes up our planet, in water and oceans, and in our building materials and homes. There is nowhere on Earth where you cannot find natural radioactivity. Radioactive elements are often called radioactive isotopes or radionuclides or just nuclides. There are over 1,500 different radioactive nuclides.

If we focus for instance on radioactivity in water, radioactive substances in ground water, such as radium, uranium and thorium, occur naturally. They are present at least to some extent in almost all rocks and radium, in particular, dissolves more readily into ground water in contact with sands or soils. Radioactivity in drinking water is not a new phenomenon, having been present to some extent for thousands of years. Nevertheless, exposure to radium over a long period of time is believed to increase one's lifetime risk of developing certain types of cancer.

Another example of radioactivity in our everydays's life is the one found in buildings. The building industry uses large amounts of waste products from other industries. In recent years, there is a growing tendency to use new construction materials with naturally or technologically enhanced levels of radioactivity (e.g. phosphogypsum, coal fly ash, oil shale ash, some rare minerals, etc.). Therefore, most building materials contain naturally occurring radioactive elements, the most important of which are ^{40}K and members of two natural radioactive series, which can be represented by the isotopes ^{232}Th and ^{226}Ra. The presence of these radioisotopes in materials causes external exposure to the people who live in the building. ^{226}Ra and ^{232}Th can also increase the concentration of ^{222}Rn and ^{220}Rn and of its daughters in the building. ^{40}K and part of the above-mentioned radionuclides cause external exposure while the inhalation of ^{222}Rn, ^{220}Rn and their short-lived progeny leads to internal exposure of the respiratory tract to alpha particles.

Increased interest in measuring radionuclides and radon concentration in building materials is due to the health hazards and environmental pollution. In view of this, there is a need to develop and introduce in both international and national levels environmentally safe and economically reasonable standard regulations, which should be based on justified radiological, social and economical legislation concepts.

In the present book, written by Prof. Fokion K. Vosniakos, M.Sc., Ph.D, a well known expert at the international level, the question of Radioactive Transfer in Environment and Food is addressed, with a completeness that is the result of decades of research in that field.

The mechanisms of radioactive elements transfer in the atmosphere, tropospheric and stratospheric diffusion of radioactivity, environmental contamination from accidents, and the impact of atmospheric pollution on the food chain—soil, plants, food are analysed. Then, the methods for analyses of atmospheric radioactivity, for analysis of the impact of atmospheric contamination on the food chain (soil, plants, and foods) are illustrated. Studies on atmospheric and environmental radioactivity, and in particular of atmospheric radioactivity in soil and plants, and of radioisotopes in food stuffs constitute the core of the book. The question of natural radioactivity concentration in building materials is addressed too, while the book contains many case studies with application and useful data for Greece, however being of general value.

However, detailed and technically up to date, with a high scientific content, the book is useful at didactical level too, remaining a complete tool for addressing the complex question of radioactive transfer into the environment and food.

Massimo Zucchetti
Politecnico di Torino, Italy
MIT Cambridge, USA

Preface

"***Every falling raindrop and snowflake carries some radioactive matter to the earth, while every leaf and blade of gross is covered with an invisible film of radioactive material***". These words, which were written by Rutherford in 1905, give a fairly realistic description of the environment from the point of view of ionizing radiation.

Several years after the Three Mile Island accident in the United States, the Chernobyl accident completely changed the public's perception of nuclear risk. While the first accident provided the impetus to develop new research programs on nuclear safety, the second, with its human death toll and the dispersion of a large part of the reactor core into the environment, raised a large number of problems of management not only for the treatment of severely exposed persons, but also for the decisions that had to be taken affecting the population. Clearly, the national authorities were not ready to manage an accident whose consequences were not confined to their territory.

During the Chernobyl accident large areas of semi-natural ecosystems were affected by radionuclide deposition. Meadows and forests are typical semi-natural ecosystems. Meadows are used extensively in many countries as pastures for cattle, sheep and goats, while forests are important to man since they provide wood, paper, wild berries, mushrooms, game and recreational areas. Post-Chernobyl investigations have shown that dose to man from semi-natural ecosystems are relatively greater than from agricultural systems and that this dose risk persist for the long-term.

Fifteen years later, many improvements in radiation protection and emergency preparedness have been made possible by the Chernobyl experience and we are also to arrive at a more accurate assessment of the impact of this accident. The fact remains that the future consequences in terms of health effects remain imprecise for simple technical reasons and because of this, lend themselves to a competition between those who want to minimize the consequences of the accident and those who which to promote a catastrophic assessment.

Studies carried out since the Chernobyl accident has increased the understanding of radionuclide behavior in semi-natural ecosystems, especially for boreal forests and middle European meadow systems, which have been extensively investigated. Data sets have been obtained which describe the distribution and the cycling of radionuclides (especially ^{137}Cs and ^{90}Sr) within these systems.

A considerable amount of knowledge on the behavior of radionuclides in the environment and many different models describing such behavior were available of the time of the Chernobyl accident. Undoubtedly, in the post-Chernobyl situation radioecology is in a better position because the description of the environment is presently much closed to reality and its conclusions much more reliable.

Therefore, the subject of environment radioactivity has aspects of vast dimensions and the task of bringing together the pertinent information in so many diverse disciplines proved to be not without its difficulties. The text should be concerned primarily with the behavior of radioactive. The important and elaborate technology by which passage of radioactive materials to the environment may be prevented and the equally important field of health physics that is concerned with protecting the atomic energy worker were thus placed beyond the bounds of this work.

I am greatly indebted to my many associates, past and present, who participated and contributed in the research of this work and especially: Prof. Dr. P. Karakoltsidis, Prof. Dr. A. Giouvanoudi, Ass. Prof. K. Farmakis, Ass. Prof. A. Kalfa, Ass. Prof. Dr. E. Sossidou, Prof. Dr. N. Zoumakis, Ass. Prof. Dr. T. Batalas, Dr. C. Hatzioannou, Ass. Prof. A. Moumtzis, Ass. Prof. Dr. P. Mentzelou, Mrs. B. Konincckx and Mrs L. Mavroudi.

Thessaloniki, Greece, 2012 Fokion K. Vosniakos

Contents

Introduction

Radioactivity, its character, intensity and impact on the planet life is a matter of prime significance. The natural atmospheric radioactivity from cosmic radiation and radioactive gases, released from the ground surface present a continuous radioactive back-ground of the earth. Two types of radionuclides can be found in the environment, those of anthropic origin and the natural radioisotopes. The natural radionuclides are essentially the cosmonuclides formed in the upper atmosphere by the interaction of cosmic ways with the elements present in the atmosphere, such as ^{14}C. ^{3}H, ^{10}B, ^{7}Be, ^{26}Al, and the primordial radionuclides formed during nucleosynthesis, which have a half- life long enough to persist on earth; their radioactive decay gives rise to decay products with shorter half- lives; these are ^{238}U, ^{232}Th with their decay series.

During the last 25 years new artificial nature contaminants appeared. They are related to the nuclear power exploitation for peaceful and war purposes, nuclear power stations and satellites, accidents in the nuclear power stations and, operation of nuclear power stations, which release many radionuclides in spite of the high standard protection systems.

Due to the Chernobyl accident (1986), vast parts of semi-natural ecosystems, such as pastures, fruit and vegetable areas, fertile lands, etc., were damaged by radionuclide deposition.

All the above mentioned bring dangerous biological consequences and poses the problem for analysis of the status and the search for ways and means for reducing the radionuclide contamination danger.

The international scientific community replied immediately to that essential necessity. Many specialized stations and laboratories were established around the world. The establishment of state, regional and international structures, aimed at the coordinated efforts of specialists to solve the problem through working out projects, research programs and educational work, etc.

In the established national and world policy, which concerns the ecological projects, there is already a considerable amount of information available about the distribution of radioactivity and transfer of radionuclides. It contributed significantly to the improvement of the radioactive protection globally.

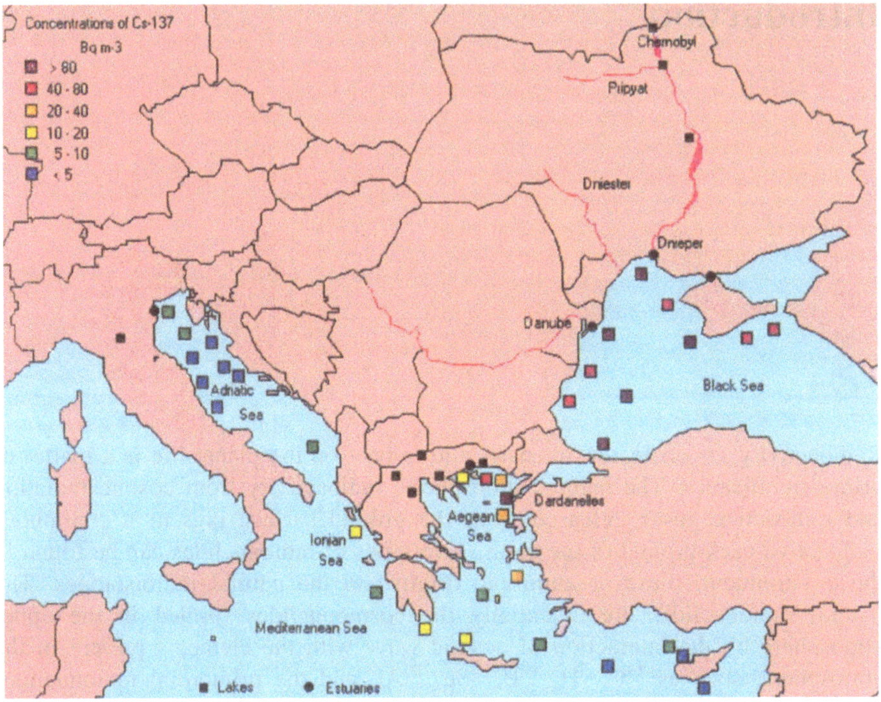

Fig. 1 Cs $^{-137}$ levels in the areas of interest during 1992–94 (Adriatic Sea 1990) [237]

Environmental radioactivity, however, has aspects of unlimited dimensions. Its explanation requires a great deal of research work by specialists of different scientific fields and the synergy of all of them. The stimulation of such an effort worldwide will enlarge the information about radioactivity and will contribute at the same time to the protection of planet life.

Nuclear power plant can cause radioactive nuclide releases over long distances from the area of the accident. The Chernobyl accident released radionuclides that reached Greece through the troposphere and contaminated its environment [219]. Crops that were planted during and after the accident, with relative superficial root system (5–10 cm) were contaminated through their root system. The transfer of various radionuclides form soil to plant is known to be affected by the soil characteristics and properties, such as texture, pH, exchangeable potassium and calcium, kind and amount of clays and organic matter content [220]. The possibility of fixation of Cs isotopes by geological material and soil has been the subject of previous studies [221]. For example, soils in Northern Greece have a high clay content with vermiculite (mica family) being the predominant mineral which is a well-known, potassium and cesium fixing material [221]. The existence of Cs-137 in the soil is important because of its possible transfer to cultivated plants and eventually animals and humans. It is also increases the direct exposure

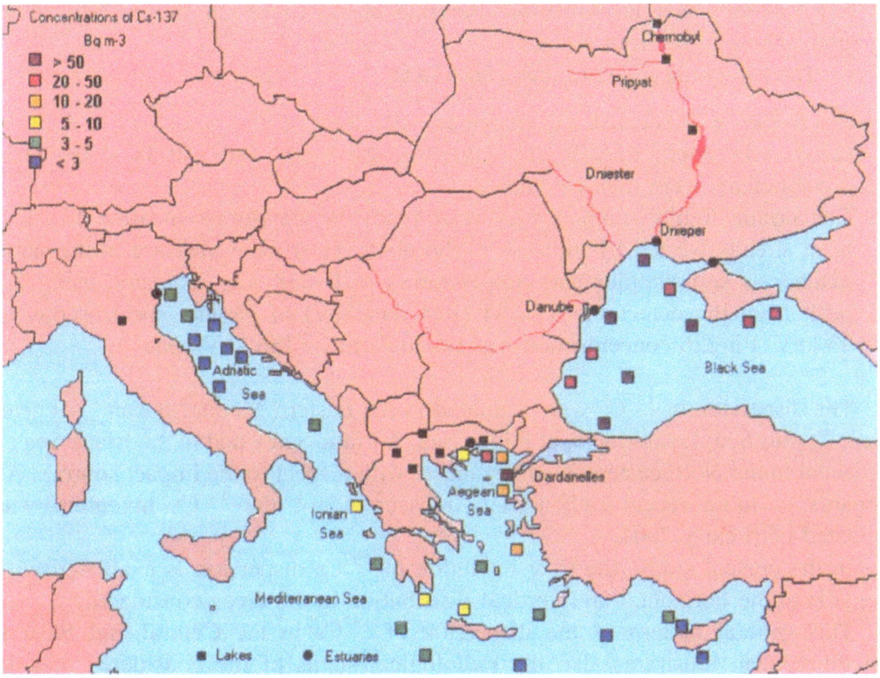

Fig. 2 Sea water Cs $^{-137}$ levels in the Eastern Mediterranean during 1997–1999 [237]

doses received by humans from terrestrial natural radioisotopes by 10% [222]. Thermoelectrical power plant operation with coal as conventional fuel is a well-known source that releases oxides that contaminate the environment. Power plants of this kind in Northern Greece use lignite as fuel that may contain natural radioisotopes of U-238 and Th-232 series (in many cases) in significant concentrations. The possibility of releasing such radionuclides in the fly coal ash has not been thoroughly investigated in the past. The aim of this work was also to compare the Chernobyl ^{137}Cs uptake from soil to corn plant in 1998.

The UNSCEAR (United Nations Scientific Committee on the Effects of Atomic Radiation) semi-empirical model was used to access contamination of plants via their root system, according to the relation:

$$C_m = pF_o \exp(-\mu m) \text{ where } m > \ = 1$$

C_m is the Cs-137 concentration in the product at the year in after deposition;
F_o (mCi/km^2) is expressed in Ceasium units (1CU = pCi Cs-137/gK), i.e. the ratio of Cs activity in the product over concentration of K in the same product;
p is a coefficient that accounts for the transfer rate of Cs from roots to product;
$p = C_m/F_o$ when exp $(-\mu m) = 1$;

μ is a constant that accounts for the rate at which Cs in the soil becomes unavailable to the root system.

The following assumptions were made:

1. exp $(-\mu m) = 1$ because C_m values calculated by other researchers in the past [223] were approximately the same for the years 1987, 1988, 1989. The p coefficients were then calculated.
2. Concerning lignite samples they were ready for measurement after they had been already treated by mill use in the plant. They were measure in the same manner as soil samples and concentration of U-238 was evaluated using the peak from the detectable Th-234 (first daughter) of gamma spectroscopy at 93 keV. Ra-226 concentration was found form its 186 keV peak.

The dispersion of ^{137}Cs was evaluated in the Eastern Mediterranean – Aegean Sea, Cyprus Sea, Levantine Sea, Ionian Sea, Adriatic Sea – and in the Black Sea as adjacent to the Northeastern Mediterranean, with respect to the impact sources. As regards to more recent radioactivity measurements, the ^{137}Cs inventory was updated until early 2001.

In the present study, the time evolution of ^{137}Cs dispersion is evaluated with regards to the horizontal and vertical distribution in the areas considered.

This general pattern of the dispersion of ^{137}Cs in the Central and Eastern Mediterranean illustrates also the radiological status of the considered area in terms of ^{137}Cs impact sources. According to this pattern, we can conclude that the impact source of ^{137}Cs is focused in the North Aegean Sea, weakening to the South. As Cs is considered as a conservative pollutant its inventory pattern is compared to other inorganic pollutants followed the same routes and dispersed to the North Aegean Sea through the straits of Dardanelles [237; Figs. 1, 2].

Chapter 1
Status of the Problem

Through radioactivity measurements, it was found that there were considerable differences in the radioactivity values in countries like Greece, especially during the periods 1961–1963 and 1969–1971. The tests showed high degree of nuclear productivity in the Balkan countries, in the former USSR, France, China and USA. It was demonstrated that the diffusion radioactivity and its transfer had increased almost two times after the Chernobyl accident [6, 8, 10, 20, 108, 184].

1.1 Radionuclides Transfer Mechanism in the Atmosphere

The release of radioactive elements may have impact to the public health in case of inhalation of the contaminated air, under extreme conditions, in case of exposure to γ-radiation from a "radioactive cloud" (i.e. a cloud with radioactive elements) or from radionuclides, deposited on the soil. In addition to this, the direct deposition of radioactive nuclides with very high activity may cause burning and other effects of the skin and the deposition in the superficial layers of plants may cause contamination to the food chain.

The atmosphere is another pathway of radioactive contamination transport to soil and waters.

To estimate the consequences from the radioactivity transfer in the atmosphere, one should take into account the amount of the released radionuclides in terms of time and location. The possibility of radiation depends on the way of its diffusion and transfer in the atmosphere, as well as on the mechanisms through which the radioactivity will be deposited on the surface of the earth. The atmosphere transports actually the radionuclides from the production sources to soil and waters [3, 25, 88, 132, 168].

In order to calculate the dose in the atmosphere from any kind of radioactivity, including the accidents cases, it is needed to be known:

F. K. Vosniakos, *Radioactivity Transfer in Environment and Food*,
Environmental Science and Engineering, DOI: 10.1007/978-3-642-28741-1_1,

- The amount of the γ-radiation dose from the "radioactive cloud"
- The amount of the γ-radiation dose from a given deposition in the soil and subsequently to the food chain
- Thyroid dose due to ^{131}I
- The amount of the lungs dose from inhalation from the "radioactive cloud".

Regarding the food chain, meaning soil, plants, foods, the following products are the most dangerous for the public health due to their contamination from radioactivity:

- *Milk*: Milk and dairy products are important food components in many countries. Milk is one of the few foods, which is produced by the animals in large areas and on a daily basis. Its consumption is almost the same globally and the sampling of liquid or dry form is very easy [121, 270].

  The first days of radionuclides release due to any kind of nuclear explosion, milk is contaminated from ^{131}I and/or $^{137,134}Cs$. Milk is mainly contaminated, when cows feeded in open areas (pastures, etc.) during the period of the radionuclides fallout to the surface of the earth. It can also be contaminated from radionuclide inhalations by the animals or from drinking water and contaminated food which is consumed by the animals. Cow milk, and especially sheep milk, (since sheep graze more than cows in open areas), should be periodically checked for a long period of time [121, 186, 270].
- *Waters (fresh and marine)*: It is well known that rainwater and snow are important environmental indicators of radioactive contamination. Drinking water and rainwater could be critical pathways of radionuclides, transfer. Water in general, drinking and using in households, are also important radioactivity pathways, through direct usage or indirect through its usage for food preparation. The water using in livestock and for irrigation of agricultural production could be a critical and significant source of radioactive contamination of the food chain. Marine water could also be a threat for the sea-foods (such as mussels, shellfish, fish etc.). River and lake waters are also considered as possible important contamination source for the environment and the food chain [270].
- *Aquatic organisms*: If we identify in a lake fish radioactive contamination, we could define the most critical especially way of the cesium absorption by the human beings. Fish in the oceans absorb less radioactive cesium than fish in fresh waters. This is due to the dilution of the $^{134,137}Cs$ in the ocean and the important dilution of ^{40}K concentration in the water. Since mussels and some species of macro-algae are quickly absorbed the radionuclides from the marine water they can be used as bio-indicators.

  The velocity of the ^{137}Cs outflow for the 0–50 m depth interval in the Black Sea, estimated on the basis of measured concentration of the period 1987–1993 has been assessed by the exponential function [290].

$$v = 25.5e^{-0.103t}$$

where v = velocity of ^{137}Cs (in T Bq/year); t = time period (in years).

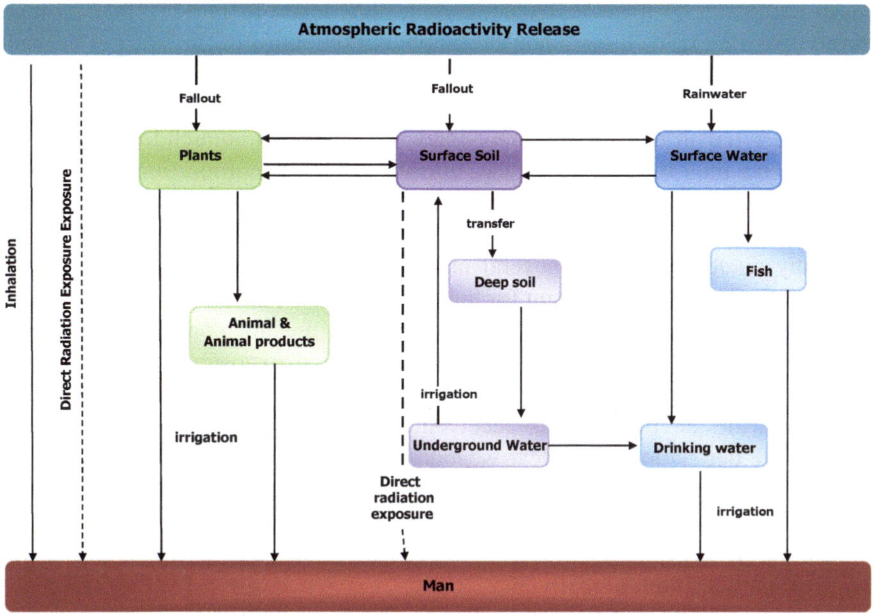

Fig. 1.1 Main pathways of radioactivity to man due to atmospheric radioactive release and its fallout

- *Atmospheric air*: The measurements of the airborne radioactivity identified the radionuclides variety concentration which causes the radioactive contamination of the atmospheric air. These radioisotopes in the atmosphere are measured in the air near to the surface of the earth. Human beings are exposed to radioactivity from the inhalation or absorption of the radionuclides, which are deposited in vegetation, or from eating animals and animal products, coming from animals which had been exposed to radioactivity substances from inhalation or absorption [8, 60, 68, 270] (Fig. 1.1).

1.1.1 Fission Products

Nuclear weapons tests in the open atmosphere were the main source of radionuclides released in the atmosphere, which were produced from the fission. During 1950–1990, people all over the earth have been exposed to these or other radionuclides. The atmospheric nuclear testing has considerably dropped off during the last two decades and nowadays it is no longer a threat for the planet and the humanity (Strategic Arms Limitation Treaty—SALT I, II, Strategic Arms Reduction Treaty—START I, II). Therefore, the major source of fission products in the atmosphere for the last twenty years has been only from the nuclear accidents (mainly Chernobyl 1986 and Fukushima 2011).

An explosion of a nuclear reactor of a NPP leads to radionuclide radiation variety like the explosion of a nuclear bomb, but the ratios of the isotopes are quite different [44, 52, 54]. The difference in the ratio between the radionuclides is due to the fact that during the reactor's operation, the concentration of radionuclides with longer half-life increases more than that of the radionuclides with shorter half-life and thus reaches to an equilibrium state. The proportionality between various radionuclides, which are produced from a NPP, changes with the time of the reactor operation and the usage of fuel.

In any nuclear explosion several hundred of radioisotopes are produced, but only a few of them are important for the public health. These would include fission and activation products. Radioactive noble gases are not considered significant since they are unlikely to contribute to the internal exposure through the food chain [182].

The marine environment is received man-made radionuclides from many sources: the detonation of nuclear weapons in the atmosphere, the controlled discharges of low-level liquid effluents from the nuclear power industry, and fallout arising from accidents such as the Chernobyl reactor accident, in 1986 and Fukushima in 2011. These various sources have, collectively, introduced a wide range of nuclides both globally and locally.

The overall impact of the releases especially from the EU nuclear installations into the Mediterranean Sea (in the period 1980–1991) was assessed by calculating the collective dose truncated at 500 years and collective dose rates to the EU population. The impact of liquid effluents discharged into the Mediterranean Sea by EU nuclear plants was found to be close to that due to ^{137}Cs present in the Chernobyl accident. The total collective dose was 1.96 man Sv, while the maximum collective dose rate was calculated to be around 0.18 man Sv y^{-1}. The contribution of the nuclear bomb fallout and the Chernobyl accident to the collective dose was estimate to be 3.7 men Sv. The highest concentrations of ^{137}Cs (within the same order of magnitude than else where) have been detected in the Aegean waters due to the inflow from the Black Sea which is influenced by the discharge from the Chernobyl area [278, 281, 283, 288, 290].

The UNEP Specially Protected Areas Protocol covers a total of 123 areas, of which 47 are marine (15 are exclusively marine and 32 are mixed land and sea). The remainder is terrestrial and may include some coastal wetland. The Specially Protected Areas Regional Activity Center of UNEP and with the collaboration of the World Conservation Union (IUCN) have listed 233 sites of biological and ecological value in a directory of marine and coastal protected areas of the Mediterranean region. These protected areas are all established under national jurisdiction, and therefore, there are few international legal instruments governing their establishment, regulation and management. The SPA Protocol to the Barcelona Convention is the principal one in the Mediterranean.

In particular [237]:

1. In the Mediterranean region the radioactivity releases are estimated to be about 5 TBq/a and less than 1 GBq/a respectively in terms of the Cs^{137} discharge.

2. Atmospheric, rivers and strait-exchange inputs Cs^{137} into the Mediterranean Sea by 1985 are estimated to be respectively, 10 ± 2 PBq, 0.4 ± 0.1 PBq which add up to the total input of 12 ± 2 PBq in 1985 for the entire Mediterranean Sea.
3. The total Mediterranean inventory of Cs^{137} (in 1985) was 11 ± 1 PBq (3.5 m Bq/l for the water).
4. The average values for sediment and biological concentration factor vary between 1 and 10^5.
5. Chernobyl fallout increases the Cs^{137} deposition about 25–40% in addition to the existing quantity from 1986.
6. Significant increases in the Cs^{137} levels in the coastal sediment of France, by a factor of 2 to 4 (on the existing Cs^{137} from 1985) has been observed.
7. The impact of artificial radionuclides on the living marine organisms in Mediterranean is negligible. The radiation risk for man may correspond of severe harm in 10^5.
8. The risk from the sea-food consumption, due to radioactivity, estimated to be about 5 cases of severe harm in 10^7 (Table 1.1).

1.1.2 Atmospheric Natural Radioactivity

Henry Becquerel in 1896 discovered the decay phenomenon and the natural radioactivity which developed later by M. and J. Curie, and F. Zoliot (1898 and 1934) and other scientists who contribute significantly on nuclear sciences and technology.

Soon it became obvious that the natural radioactivity was a useful tool, which allowed a more accurate study of the structure and properties of the matter, especially microscopically. Natural radioactivity has been used, for example, to identify the structure of the atomic nucleus, to evaluate the age of the earth, etc. [32, 52, 107].

Nowadays the peaceful use of nuclear energy, as electrical production, makes the study of all kind of radioactivity necessary in order to understand the environmental impact of the radioactivity produced by the NPP. Only by having the necessary information about the amounts of radioactivity produced during a normal NPP operation and during a possible accident and its impact to the environment we could interpret the monitoring in the neighboring locations of NPP and the transfer of it to marine, to atmosphere and to human beings.

Small amounts of natural radioactivity can be identified in any alive or non-alive material. Extraterrestrial ionizing radiations continually hits the earth and subsequently all human beings. This is a permanent exposure to such radiations for the inhabitants of the planet (human and animals). Therefore the natural radioactivity, to which all the earth's inhabitants are exposed, is the combination of terrestrial and extraterrestrial radiation.

Table 1.1 Fission and activation products due to a NPP accident [285]

	Nuclide	Half-life	Fission yield (%)	Major decay
Fission products	Sr- 89	50.5 d	4.77	β^-
	Sr-90, Y-90	28.7 a, 64.1 h	5.76	β^-, β^-
	Zr-95, Nb-95	64.09 d, 35.0 d	6.51	$\beta^-\ \gamma$, $\beta^-\ \gamma$
	Mo-99, Tc-99m	2.747 d, 6.006 h	6.09	$\beta^-\ \gamma$, $\beta^-\ \gamma$
	Ru-103, Rh-103m	39.272 d, 56, 116 min	3.03	$\beta^-\ \gamma$, $\beta^-\ \gamma$
	Ru-106, Rh-106	372.6 d, 29.92 s	0.4	β^-, $\beta^-\ \gamma$
	Te-129m	33.6 d	0.661	$\beta^-\ \gamma$
	I-131	8.021 d	2.875	$\beta^-\ \gamma$
	Te-132, I-132	76.856 h, 2.3 h	4.282	$\beta^-\ \gamma$, $\beta^-\ \gamma$
	Cs-137, Ba-137 m	30.0 a, 2.55 min	6.136	β^-, γ
	Ba-140, La-140	12.751 d, 1.6779 d	6.134	$\beta^-\ \gamma$, $\beta^-\ \gamma$
Activation products	Ce-144, Pr-144	284.45 d, 17.25 d	5.443	$\beta^-\ \gamma$, $\beta^-\ \gamma$
	H-3	12.35 a		β^-
	C-14	5730 a		β^-
	Fe-55	2.75 a		EC
	Fe-59	44.53 d		$\beta^-\ \gamma$
	Mn-54	312.5 d		EC, γ
	Co-60	5.27 a		$\beta^-\ \gamma$
	Zn-65	243.9 d		EC, γ
	Cs-134	754.2 d		$\beta^-\ \gamma$
	Np-239	2.355 d		$\beta^-\ \gamma$
	Pu-241, Am-241	14.35 a, 432.0 a		β^-, $\alpha\gamma$
	Cm-242	162.94 d		α
	Pu-238	87.7 a		α
	Pu-239	2.411×10^4 a		α
	Pu-240	6.563×10^3 a		α
	Pu-242	3.735×10^3 a		α

About 340 elements have been found in nature, of which about 70 are radionuclides belonging to the heavy elements category. Elements with atomic number (Z) greater than 80, have radioisotopes. The relative abundance of the isotopes on earth is derived from the isotopic ratios produced since the universe creation. In 1950s it was announced that the ratio of the elements was not only on earth but everywhere (in the atmosphere, in the meteorites, and even in the extraterrestrial substances).

It has been identified that the constant isotopic distribution is a basic property of the universe. Nevertheless, slight declination of the constant isotopic ratios may be happening due to the physicochemical processes occurring geologically.

Soil natural radioactivity presents great variations, almost everywhere, due to the different radionuclides concentration and also due to the soil composition (for example, soil with high organic matter deposit big quantity of radioactivity fallout and vice versa, as it is the case for greek soil).

Table 1.2 Residual radioactive material in the global environment as a result of the Chernobyl accident [237]

Significant radionuclide	Released in 1986 ($\times 10^{15}$ Bq)	Remain in 1996 ($\times 10^{15}$ Bq)	Remain in 2056 ($\times 10^{15}$ Bq)
I-131	1200–1700	0	0
Sr-90	8	6	1.5
Cs-134	44–48	1.6	0
Cs-137	74–85	68	17
Pu-238	0.03	0.03	0.02
Pu-239	0.03	0.03	0.03
Pu-240	0.044	0.044	0.03
Pu-241	5.9	3.6	0.2
Pu-241[a]	0.005	0.08	0.2

[a] The activity of Am-241 in 1990 has increased since 1986, as it is a daughter product of Pu-241

Table 1.3 Activity concentrations of natural radionuclides in sea water worldwide (mBq I^{-1}) [237]

Nuclide	Concentration $\times$ 10^{-6} (mBq I^{-1})	References	Region
^{238}U	47	Bojanowski et al. [271]	Mediterranean Sea
	44.4	Miyake et al. [272]	North Pacific Ocean
^{236}Ra	1.48–1.67	Broecker and Cromwell [273]	Atlantic Ocean
	1.20	Szabo [274]	Indian Ocean
		Bojanowski et al. [271]	Mediterranean Sea
^{232}Th	(3.7–288.6) $\times$ 10^{-4}	Miyake et al. [272]	North Sea
		Moore and Sacket [275]	Atlantic Ocean
		Somayajulu and Goldberg [277]	
^{40}K	11,840	Eisenbud [52]	North Sea
^{228}Ra	(3.7–37) $\times$ 10^{-2}	Kaufman [276]	Baltic Sea
^{228}Th	0.007–0.12	Miyake et al. [272]	North Pacific Ocean
		Moore and Sacket [275]	Mediterranean Sea
		Bojanowski et al. [271]	

The natural radionuclides are divided into three families of radionuclides (1) the uranium family with parent isotope, ^{238}U; (2) the thorium family with parent isotope, ^{232}Th and (3) the actinium family with parent isotope, ^{235}U.

Those three families of natural radionuclides can be found on the earth's crust and they represent the main natural radioactivity to which the human beings and animals are exposed. A fourth family, the neptunium family, with parent isotope, ^{241}Pu, also exists, but is of minor importance due to its short life-time, since ^{241}Pu has a half-life of only 14-years [1, 126, 138, 140, 160] (Tables 1.2, 1.3, 1.4).

Table 1.4 Technical specification of the nuclear power plants in operation in countries bordering the Mediterranean Sea and Black Sea [237, 278, 279]

Country	Reactor type/Name (number of units)	Net capacity (MWe)
France	PWR Bugey (4)	4 × 920
	PWR Cruas-Meysse (4)	3 × 880, 1 × 915
	PWR/St Alban (2)	2 × 1335
	PWR/Tricastin (4)	4 × 915
Spain	PWR/Asco (2)	2 × 887
	BWR/Cofrentes (1)	994
	BWR/Garona (1)	460
	PWR/Vandellos (1)	943
Slovenia-Croatia	PWR/Videm-Krsko (1)	632
Bulgaria	VVER/Kozloduy (2)	2 × 953
Russia	VVER/Balakovo (3)	3 × 950
	RMBK/Kursk (4)	4 × 925
	VVER/Novovoronezh (3)	2 × 385, 950
	RMBK/Smolensk (3)	3 × 925
Ucraine	RMBK/Chernobyl (2)	2 × 925
	VVER/Khmelnitski (1)	950
	VVER/Rovno (3)	361, 384, 950
	VVER/South-Ucraine (3)	3 × 950
	VVER/Zaporozhe (5)	5 × 950

1.1.3 Environmental and Food Radioactivity

Regarding the internal contamination of human beings from the consumption of food and water with radionuclides, which are part of the immediate pathways leading to contamination of food, the most important radionuclides which have to be investigated, following a radioactive release from a reactor of NPP, in the environment are: ^{134}Cs, ^{137}Cs, ^{137m}Ba, ^{131}I and other gamma emitters, the beta emitters ^{89}Sr, ^{90}Sr and tritium, and the alpha emitters with ^{238}Pu, $^{239+240}$Pu, ^{241}Am and ^{242}Cm [53, 192, 270].

According to relating research by UNSCEAR [270], the significant radionuclides for the contamination of the foodchain and environment are:

Isotope	Air	Drinking water	Marine–fresh water	Soil	Vegetation	Milk	Meat	Other foods
^{3}H		●						
^{59}Mn			●					
^{55,59}Fe			●					
^{60}Co			●					
^{65}Zn			●					
^{89,90}Sr		●		● (^{90}Sr)	●		●	●

(continued)

(continued)

Isotope	Air	Drinking water	Marine–fresh water	Soil	Vegetation	Milk	Meat	Other foods
^{95}Zr			●		●			
^{95}Nb			●		●			
^{103,106}Ru			●		●			
^{110m}Ag			●					
^{125}Sb			●					
^{131}I	●	●	●		●	●		●
^{134,137}Cs	●	●	●	●	●	●	●	
^{141,144}Ce			●		●			
^{238,239,240}Pu				●				
^{241}Am				●				
^{242}Cm				●				

The aforementioned radionuclides are the most important isotopes concerning the foods production. In fresh water and marine eco-systems the biological processes could result in transfer and enrichment of specific radionuclides very rapidly. Plankton and algae are capable to accumulate the radionuclides, which enter to their ecosystem. With this way the radionuclides could concentrate in aquatic organisms (oysters, clams, shrimps, etc.) [81, 82, 192, 270].

A variety of radionuclides can be present in the location of a nuclear accident from a NPP and their potential contribution to human and other living resources exposure depends on the type of accident (according to INES-scale) and the circumstances at the time of the nuclear accident.

1.1.4 Dose from a "Radioactive Cloud" and from Inhalation

The calculation of the total dose of radioloactivity from a certain source at any point on wind's direction is a complex issue since it depends mainly on the meteorological parameters. It is necessary to take into account all the effects, such as cloud dimensions and variety of fission products, involved in both dose and soil concentration. Approximations can be applied due to the great uncertainties involved in any dose measurements for a specific location and conditions.

1.1.4.1 γ-Radiation Dose from a "Radioactive Cloud"

Many accuraty solutions to the γ-radiation dose from radioactive cloud may be found. A simple approach to this issue is based on the assumption that the human beings are on the ground in an infinite medium through which radioactivity is uniformly dispersed during the period in which one could receive the total dose (Bq/m^3).

The monitoring dose is easier to be calculated by neglecting any backscattering radiation from the environment. This approach which reduces the total dose exposure can be confronted by the acceptance of the radioactive cloud to be with infinite dimensions [1, 20, 52, 67, 184].

The dose from the "radioactive cloud", containing 3.7 × 10^{10} Bq/m^3 (1 Ci/m^3) is given by [52]:

$$D = \frac{1}{2}(\overline{xt})(\overline{E}_\gamma)\frac{1.6 \times 10^{-6} \times 10^{-7}\,\mathrm{J/MeV}}{\left(\frac{10^2 \times 10^{-7}}{10^{-3} \times 10^{-2}}\,\mathrm{J/kgGy}\right)\left(0.0012\,\frac{10^{-3}}{10^{-6}}\frac{\mathrm{kg}}{\mathrm{m}^3}\right)}$$

or,

$$D = \frac{1}{2}(\overline{xt})(\overline{E}_\gamma)\frac{1.6 \times 10^{-13}\mathrm{J/MeV}}{\left(1\frac{\mathrm{J}}{\mathrm{kgGy}}\right)\left(1.2\frac{\mathrm{kg}}{\mathrm{m}^3}\right)}$$

and finally,

$$D = 0.67(\overline{xt})(\overline{E}_\gamma)\left[\mathrm{Gy/m^3/MeV}\right]$$

where D = dose (Gy); $\overline{E}_\gamma$ = mean energy (MeV); $\overline{xt}$ = average product of concentration multiple by the time (Bq s/m^3).

Therefore the dose of a "radioactive cloud" with contamination 1 Ci/m^3 is given by the above relation in Gy. This dose is not necesery to be inhelated by the inhabitants. The effective dose to man is much lower than this D.

1.1.4.2 Inhalation Dose of the Lung

Assuming that the time of radiation from any radioisotope releases in the atmosphere due to any reason (normal NPP operation, nuclear accident etc.) and human exposure is shorter that the half life of the radionuclide releases then the lungs dose could be calculated by [52] for dose $\overline{xt}$ (Bq/m^3):

$$D_L = (\overline{xt})\,(E_\gamma)\frac{\left(115 \times 10^{-6}\mathrm{m^3/s}\right)(0.12)\,(24 \times 3600\,\mathrm{s})\,\left(1.6 \times 10^{-7}\,\mathrm{J/MeV}\right)}{10^2\,\mathrm{M}}(\mathrm{RBE}) \times \int_0^t exp[-0.693\,T_r t]dt$$

Finally,

$$D_L = \frac{\overline{xt}}{MT_r}\mathrm{E}_\gamma\,2,76 \times 10^{-5}[1 - \exp(-0.693T_r t]\ (\mathrm{RBE})$$

where: D_L = lung dose (in Sv) for time t; $\overline{xt}$ = exposure (Bq/m^3); 115 × 10^6 m^3/s = volume of air inhaled per given unit of time; 0.12 = fraction of inhaled dust retained in lower lungs at the end of 1 day (24 × 3600 s); M = mass of lungs (kg); E_γ = energy of particulate radiation (MeV); RBE = relative biological effect; t = days; T_r = relative half-life in days

$$\left[T_r = \frac{T_n T_b}{T_n + T_b} \quad \begin{array}{l} \text{where } T_n = \text{natural half life} \\ \text{and } T_b = \text{biological half life} \end{array}\right]$$

1.1.4.3 Iodine Dose of the Thyroid

The meltdown of a reactor from any NPP contaminates the atmosphere as a function of time regarding the portion of radioactive release. In the beginning of a nuclear accident (such as Chernobyl or Fukushima) the first radioisotope which is realeased is the iodine 131. According to the selective rules, for iodine, the thyroid is the gland where this radioisotope will be concentrated.

If the dose is expressed in Bq, for the ^{131}I the thyroid dose can be estimated from the following relation [52]:

1 Ci/m^3 = 330 rem or by substitution, by the I.S. units, we obtain: 3.7×10^{10} Bq/m^3 = (330/100) Sv, which finally is: 1 Bq/m^3 = 0.89×10^{-10} Sv = 0.089 nSv.

1.1.5 Atmospheric Dispersion of Radioactivity

In the case of a nuclear accident from NPPs or from any other nuclear and thermonuclear explosions, the radioactivity which is released is spread through atmosphere all over the world.

The dispersion of any substance in stratospheric part of the earth atmosphere was based on the dust samples behavior obtained by meteorological or balloons in an altitude of about 40,000 m. The distribution of ozone and water vapor and of a wide variety of radioactive substances including radioactivity from nuclear explosions have been examined by many scientist. The transport of natural radioisotopes like ^{7}Be, ^{32}P, and ^{14}C, components of the cosmic ray, has also been studied in detail. The cloud from a certain nuclear test was identified to about 43 km in Pacific while another's traces reached to 100 km. In 1962, an explosion which have been observed in about 400 km in Pacific contained an amount of ^{109}Cd, which was measured in many places around the world.

In 1950s it has been estimated that the lower stratosphere has a greater degree of turbulence than previously believed mainly due to the results of studies of substances coming from thermonuclear explosions [182].

Brewer and Dobson [15] realized significant stratospheric measurements on the kinetics of the stratospheric water and ozone. Stewert based on the results developed a model of stratospheric–tropospheric exchange.

This model, is based to the assumption that air enters the stratosphere in the tropic regions where it is heated and rises to an altitude of about 30 km where it begins to transfer downwards, in the direction to the poles.

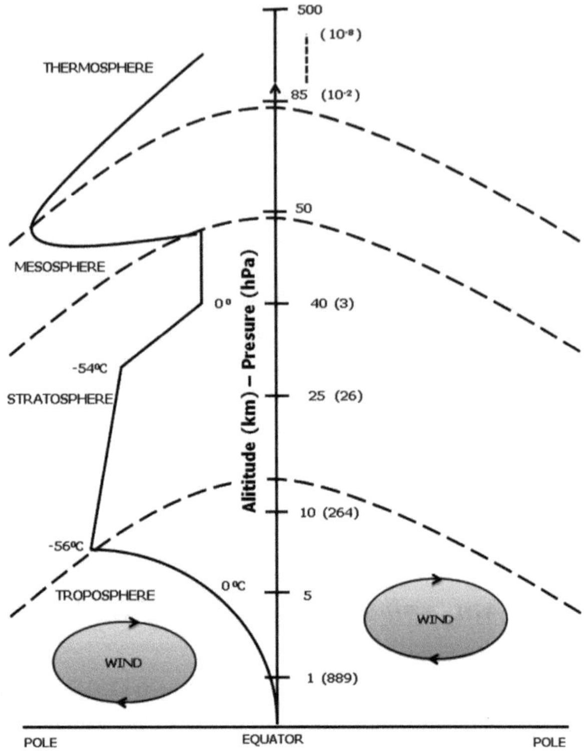

Fig. 1.2 Earth atmosphere and wind turbulences

As shown in Fig. 1.2, the troposhere is lower in the polar regions than at the equator, and its discontinuities in the temperate regions facilitate transfer from the stratosphere to the troposphere. The jet streams from the west occur these discontinuities with velocities of 100–300 km/h. The transfer rate from the lower stratosphere is more rapid in the winter and early spring.

Radioactivity stays in the stratosphere and its time depends on the altitude, the period of year, and of course the latitude. Radioactivities in the lower polar stratosphere produced by any thermonuclear explosion and remain there in average less than 5 months, whereas in tropical latitudes this time is between 2–3 years for the middle stratosphere and 5–10 years for the beginning of thermosphere. Diferent aerosols in the troposphere are distributed all over and they are deposited on the surface of the earth, mainly by the rain-falls. The time of stay for any substance into the troposphere is on average about 30 days. This time can be varying from 5 days for the lower rain-bearing region of the troposphere to 40 days in the higher altitudes. In fact, rainfalls retransfer any aerosol from the troposphere (rain-out) and also by wash-out [22, 52, 57].

It is believed that ocean spray is effective in the substances distribution close to the ocean-atmosphere border, and may be this fact might explain that the fallout of ^{90}Sr into the oceans is higher than on land [57].

1.2 Environmental Impact from Nuclear Accidents

When a nuclear accident occurs, in which radioactive substances are released to the environment, there is always a great opportunity to collect significant information regarding the way the release took place, the mechanisms through which the radioactivity was distribuded into the environment, and its impact on the ecosystem and public health.

1.2.1 Information from the Fallout of a Thermonuclear Explosion

Just before March 1, 1954, very few people were informed that the fallout to the land can provoke explosion in the range of megaton which subsequently could produce significant amounts of radioactivity fallout over large areas. In March 1, 1954 any existied doubts dissapeared when extensive fallout occurred following explosion of many megatons, indentified by a ship close to Bikini Atoll in the middle Pacific Ocean.

The increase of γ-radiation detected in Rogerik Island (about 110 km east of Bikini) was the first indication of radioactivity fallout. The fallout began few hours after the explosion, and shortly it recorded a rate of about, 100 mR/h. The evacuation of the 28 inhabitants began immediately after the explosion. Rongerlap, Utrik, and Alinginae were evacuated subsequently [259, 260].

It was a few days later when it became known that a Japanese fishing vessel was about 100 km to the east of Bikini at the time of the fallout. Although the fallout was visible on both Rongelap and the Japanese fishing boat, neither the Marshalles, nor the Japanese took any precautions to minimize the amount of their exposure. This is understandable if you take into acoount their relative ignorance on the matter. Thus, the natives of Rongelap and the fishermen of the Japanese fishing vessel lived from few hours to some days due to the radioactivity exposure (atomic sickness). Even visuable radioactive materials of fallout felt on their heads and some others into direct contact with their skin. The eyewitnesses (Japanese fishermen), described the radioactive dust deposit as snow fall.

In 1960s specific researches and studies took place in order to evaluate the radioactive impact to the local environment. Through these studies it was estimated that the ^{90}Sr impact to the natives in 1957 was 0.1 nCi (37 Bq). This impact has increased to 2 nCi (74 Bq) by 1958 and 6 nCi (222 Bq) by 1959 due to the radioactivity accumulation to the environment. By 1962, it was estimated from urine analyses of natives that the average body impact of ^{90}Sr was 12 nCi (444 Bq) [1, 27]. The impact to the public health of both ^{137}Cs and ^{90}Sr started to drop continuously till 1963. It was unknown whether this drop was due to lower radioactive contamination of food or due to the fact that the daily intake has been diluted by an increasing amount of imported "clean" food [262].

1.2.2 The Accident of Chernobyl

The main targets for the designers and constructors are the reliability and safety of any nuclear power station (NPP). It is not without reason that the opponents of nuclear power have consistently highlighted problems of reliability and safety at these factories.

Realizing the vital importance of reliable operation, the designers and builders of nuclear power stations and their directors have tried to foresee all measures, which they consider essential to ensure safety. Many possible combinations of accidents have been taken in consideration and the designers have endeavored to anticipate all the steps required preventing accidents and, particularly, to ensure safe operation when problems occur in the core of the reactor [36].

Nevertheless, unexpected accidents have taken place at different times in many countries:

- In 1957 there was an accident in a reactor at the Windscale plant in the United Kingdom, which involved a release of radioactive fission products [36, 96, 97].
- In 1959 some of the nuclear fuel elements on a reactor at Santa Susanna, California, USA, melted.
- In 1961 there was an explosion in a reactor at Idaho Falls, USA.
- In 1966 there was a partial core meltdown in the Enrico Fermi reactor in Detroit, USA.
- In 1979 the core melted in one of the reactors at Three Miles Island, Pennsylvania, USA.
- In 1982 a steam generator pipe ruptured at the Ginna reactor in the United States, releasing radioactive steam to the atmosphere.

It is obvious that there is no need to list all accidents that have occurred at nuclear power stations; it is enough to say that during the period between 1971 and 1985 there have been 151 accidents of varying degrees of seriousness in 14 countries all over the world, and that these accidents had certain consequences—(sometimes quite serious consequences) for people and the environment [225].

The accident in the fourth unit of the Chernobyl (Ukraine) nuclear power station, which took place on April 26, 1986 and until today is considered as the biggest nuclear accident in the world, led to very serious consequences for the whole Europe. Thirty one people died as a result of the accident and many others have suffered serious health damage. The destruction of this RBMK-reactor (former USSR constraction) resulted in radioactive contamination of the region surrounding the NPP over an area of about 1,000 km^2. In this region the crop fields have been taken out of cultivation and work has had to be stopped in factories on building sites. Several tens of thousands of people had to evacuate a region with a radius of 30 km around the power station.

The Chernobyl accident was caused by a whole series of errors mainly because the personel of the NPP did not follow the international rules for safety in NPPs. The size of the catastrophy of the accident was due to the big release of radioactivity, compounded by staff's failure to observe that the operating rules introduced a dangerous situation unpredictable in the design calculations.

On April 25, at one o'clock in the morning, the power station staff began procedures to shut down the reactor, which until then was operating at rated power. According to the power station staff, the reactor was to be shut down for regular maintenance since at the time of shutdown the core, with 1659 fuel assemblies, had reached an average burn up for 10.3 MWd/kg.

Before shutdown, the administration of the Chernobyl power station decided to carry out tests on turbo generator no. 8 in a regime whereby the turbine would be supplying NPP power requirements during the shutdown. The tests were scheduled to be carried out during the night, but the personel in charge of the NPP did not prepare the experiment in the proper way and they did not secure the approval of the competent organizations, although they were obliged to do so.

Shortly after the nuclear accident, the main concern was concentrated on the avoidance of uptake of radioisotope ^{131}I by the thyroid gland, where it effectively concentrates, primarily through composition of milk and leafy vegetables. Because of the short half-life (approximately 8 days) of ^{131}I and therefore, the relatively short period for which relevant protective measures are needed, of many countries' authorities (especially in North and North-East Europe) introduced measures to avoid radiation exposure of thyroid gland of the inhabitants and specifically of the youngs (by providing pills of non-radioactive iodine, which is occupied the thyroid glant leaving no space for ^{131}I accumulation in the gland).

From the radioactive contamination point of view, the radioisotopes $^{134,137}Cs$ are the most persiment contamination. ^{137}Cs has relatively long half-life (approximately 30 years). It contributes to the radiation dose received by man in two ways: firstly, externally from the contamination on the ground and other surfaces and secondly, internally by consumption of contaminated food. While taken through food consumption, it is distributed throughout the soft tissues of the body (Table 1.5).

Considering the radioactivity accumulation into ecosystems it should be distinguished that after one half-life of the isotope concerned; only half amount of its radioactivity remains in the ecosystem. Similarly, after ten half-lives, 1,000 of the total radioactivity accumulated in the ecosystems still remains. For example, ^{137}Cs (with about 30 years half-life) which was produced by the Chernobyl accident in 1986, it will remain half of it till 2016 and almost 0,001 of ^{137}Cs after 300 years.

Table 1.5 Radionuclides found in the air and deposition samples, just after the nuclear accident at Chernobyl NPP [225]

Nuclide	Half-life	Major decay
H-3	12.35 a	β^-
Sr-89	50.5 d	β^-
Sr-90	28.7 d	β^-
Zr-95	64.09 d	β^- γ
Nb-95	35.0 d	β^- γ
Mo-99	2.7476 d	β^- γ
Ru-103	39.272 d	β^- γ
Ru-106	372.6 d	β^-
Ag-110 m	249.79 d	β^- γ
Cd-115	2.2 d	β^- γ
Sb-125	1008.1 d	β^- γ
Sb-127	3.9 d	β^- γ
Te-129 m	33.6 d	β^- γ
Te-131 m	30.0 d	β^- γ
Te-132	3.204 d	β^- γ
I-131	8.021 d	β^- γ
I-133	20.3 h	β^- γ
Cs-134	574.2 d	β^- γ
Cs-136	13.0 d	β^- γ
Cs-137	30.0 a	β^-
Ba-140	12.751 d	β^- γ
Ce-141	32.50 d	β^- γ
Ce-144	284.45 d	β^- γ
Np-239	2.355 d	β^- γ
Am-241	432.0 a	α γ
Cm-242	162.94 d	α
Pu-238	87.70 a	
Pu-239/240	2.411×10^4 a / 6.563×10^3 a	α/α
Pu-241	14.35 a	β^-
Pu-242	3.735×10^5 a	α

Artificial and natural radioactivity exists almost in every living and not living material. For example:

- The man has 3,000 Bq,
- 1 kg coffee has 1,000 Bq,
- 1 kg granite laminate has 1,200 Bq,
- 1 kg phosphate fertilizer has 5,000 Bq,
- A medical isotope for diagnosis has 7×10^7 Bq,
- 1 tone fly coal ash has 2×10^6 Bq,
- A source for radiotherapy has 1×10^{18} Bq,
- 1 tone of U has 1×10^{10} Bq, etc.

The sarcophagus built on the reactor no. 4 of the Chernobyl NPP was proving after all, that it was not the appropriate solution in order to protect the rest NPP from the high doses of the debris of reactor no 4. Consequently such defensive protection levels cannot improve the safety of a destroyed reactor and guaranty the secure in permanent basis. Additional practical solutions and constructions should be preferred in order to take into account the radiological risk and the safety priorities of the particular case.

Summarizing the impact of this accident we may now clearly understand that an area of about 15,000 km^2 (mostly 75% belong to Belorussia) was infected mainly by ^{137}Cs. More than 100,000 people were moved to several other places for safety reasons with the possibility to never return back to their land. A significant problem still remains, is the uncontrolled return to Chernobyl area of some inhabitants with a great risk for their life, living in such highly contaminated locations.

1.2.3 Worldwide Nuclear Power Status

The year 1985 is identified as the the year with the greatest increase in the use of nuclear power to electricity production. In fact, during 1985, the total installed nuclear power capacity in the world increased by 14%, with 32 new Nuclear Power Units (NPPs) with a total capacity of 30 GW electric (GWe) being connected to the national electrical networks. At the end of 1985 there were 374 NPPs, with a total capacity of electrical production about 250 GWe all over the world.

During 1985, NPP generated about 1,400 TWh of electricity globally, which reflected an increase of 19% over 1984, and accounted for about 15% of the world electricity generation during 1985.

The electricity produced by NPPs worldwide during 1985 was of the same order as the total electricity generated by the NPPs of the 10 Member States of the European Union in this year. In 1985, the produced electricity amounted to 1,400 TWh. In the Western Europe, the electricity production from NPPs in 1985 was 551 TWh and was the same with the total electricity production in 1960.

The nuclear share in electricity generation greatly varies from country to country, and also from region to region in some countries (USA). There were 19 countries in which their NPPs contributed 10% or more to the total electricity in the world, produced in 1985. In the countries belonging to the Organization for Economic Cooperation and Development (OECD), about 20.4% of the total electricity generated in 1985 was produced by NPPs.

It should be noted that while the contribution of NPPs to electricity in Canada in 1985 was 12.7%, in the province of Ontario was 42%. Another typical example of different nuclear power use was in the USA, where the nuclear power contributed 15.5% to the overall electricity production in the country, while in six states was over 50%.

1.3 The Impact of the Atmospheric Pollution on the Food Chain—Soil, Plant, Food

Contamination of vegetation may result from two pathways. The importance of those two pathways varies according to the time and the scenario. Direct contamination from the atmosphere prevails over routine gaseous radioactive releases (^{3}H, ^{14}C and radioiodines) and accidental releases mainly of volatile radioiodine and radiocaesium and less volatile (e.g. strontium) radionuclides. With time, radioactive material, from both routine and accidental releases, which are deposited on the ground accumulates and constitutes the source of indirect contamination of plants by root uptake, resuspension of dust and rain splash. The deposition of ^{90}Sr or ^{137}Cs in the soil is given by the following equation [52]:

$$C = p_r F_r + p_d F_d$$

where C = 12 month mean ratio of ^{90}Sr to calcium in milk (pCi/g Ca); F_r = annual deposit of ^{90}Sr (mCi/km^2/year); F_d = the cumulative deposits of ^{90}Sr (mCi/km^2); p_r = the proportionality factor for the rate-dependent component of C; p_d = the proportionality factor for the deposit-dependent component of C.

Radioactivity exists in the atmosphere as particles, aerosols or gas tends to fall out and is being deposited on the ground and on vegetation surfaces when present. The deposition velocity and filtration efficiency by the vegetation depends on the physicochemical characteristics of the plants (morphology, structure and composition of the surface), density of the vegetation cover (seasonality) and prevailing climatic conditions (wind speed, relative humidity, etc.). The total dry deposit on a plant canopy can be estimated as the product of the deposition velocity (V_d) and the mean concentration (routine releases) or the time-integrated concentration (accident) in air (C_a). Values of V_d depend on the aerosols characteristics and nature of deposition surfaces. Deposition velocities (V_d in ms^{-1}) are given as follows for soil and maximum leaf area index [237]:

Surface type	Aerosol bound radionuclides	Elemental iodine	Organic iodine
Bare soil	5×10^{-4}	30×10^{-4}	0.5×10^{-4}
Grass	15×10^{-4}	150×10^{-4}	1.5×10^{-4}
Trees	50×10^{-4}	500×10^{-4}	5×10^{-4}
Others	20×10^{-4}	200×10^{-4}	2×10^{-4}

Only a fraction(f) of the total deposit is intercepted by plants; this fraction can be represented by the following relation [284]:

$$f = 1 - e^{-\mu B}$$

where B represents the standing vegetation biomass (kg dw m^{-2}). Values reported by different authors for the interception coefficient (μ) range from 0.2 to 17, depending on the size of the deposited aerosols, on the plant species and on the humidity degree of the leave surfaces. Interception coefficient by wet leaves are 2–3 times higher than that for dry leaves. Another way to estimate the intercepted fraction is by considering a linear relationship between interception yield and Leaf Area Index (LAI) at the time of the deposit. The interception efficiency by the vegetation depends on the size of the droplets and the amount of rainfall, as well as on changes in radionuclides concentration in the rain water as a function of the length of the rainfall period. The foliar surfaces are able to retain a certain quantity of water. Therefore, the interception of radionuclides by wet deposition is ruled by the water storage capacity and surface (LAI) of the plant leaves. Of course, if rain lasts, the atmosphere is progressively washed out and less contaminated rain drops will then reach the plant and leach off part of the already deposited radioactivity down to the soil. The processes of plant contamination by sprinkled irrigation water ar similar to those mentioned for wet deposition from atmosphere. Foliar contamination is reduced by physical decay, weathering processes (wind, leaching by rain, fog, dew and mist or irrigation water), senescence processes (shedding of cuticular wax, de-back of old leaves). The radioactivity in plants is also diluted by plant growth and removal of contaminated parts by harvesting or grazing. The rates of these processes, excluding physical decay, are estimated by "field loss half-time" (T_w), also called the "environmental or ecological half-time", that is the time needed to reduce the contamination level on vegetation by a factor of 2. The combined action of environmental removal processes and of physical decay is termed "effective half-time".

The uptake of radionuclide from the soil by plant roots varies widely, depending on the plant species, the environmental conditions, and the soil characteristics. The accumulation in the whole plant or plant parts relative to the contamination level in soil, is quantified by the soil-to-plant transfer factor (B_v) which represents the ratio of the radionuclide content in the plant (Bq g^{-1} dw or fw) to the contamination level in the soil (Bq g^{-1} dw). Measurements carried out after Chernobyl in different regions in Greece and for different plant species have provided B_v for ^{137}Cs is given as follows [139]:

^{137}Cs transfer factor (B_v) for different plant species [139]

Plant species	B_v	Plant species	B_v	Plant species	B_v
Wheat (grain)	1.4×10^{-2}	Wheat (straw)	3.0×10^{-2}	Tomato (fruit)	7.0×10^{-2}
Rice (grain)	3.7×10^{-2}	Rice (straw)	8.6×10^{-2}	Melon (fruit)	1.6×10^{-1}
Maize (grain)	1.8×10^{-3}	Maize (green)	3.6×10^{-2}	Pear (fruit)	9.1×10^{-2}
Fava bean (Seeds)	6.3×10^{-2}	Fava bean (green)	1.3×10^{-1}	Alfalfa (hay)	2.0×10^{-1}
Soybean (seed)	1.8×10^{-2}	Soybean (green)	4.9×10^{-1}	Grass	1.3×10^{-1}
Bean (seed)	1.8×10^{-2}	Spinach (leaves)	2.5×10^{-1}	Tobacco (leaves)	5.5×10^{-1}
Chick pea (seed)	7.5×10^{-2}	Beet (leaves)	4.8×10^{-1}	Beet (root)	2.3×10^{-1}

It is known that the radioactive contamination penetrates in soils and plants and therefore they transfer to man. Most of the foods consumed by man are grown on soil. The significance on public health depends on the degree of radionuclide transfer to the food chain. The food chain depends on peoples' diet of the area concern. For example, the "Mediterranean Diet" is a well known diet internationally. The term "Mediterranean Diet" is however very general, given the large differences in the dietary habits among the different Mediterranean populations, e.g. between the Spanish, Italians, French or Moroccans. For example, Greeks eat more cheese, less butter and drink less milk compared to the French and Italians. However, beyond the apparent differences some nutritional characteristics are common to all or to the most of the Mediterranean area. The Mediterranean diet is rich in foods like legumes, cereals, fruits, vegetables and therefore rich in fibres. Olive oil is the basic source of fat. The Mediterranean's eat moderate amounts of fish, poultry, eggs, milk and milk products. Red wine is consumed in small quantities with their meals.

In Mediterranean diet the consumption of meat has increased over the past 30 years by 24% according to our data. The mean availability of meat (including beef, pork and lamb) today is 100–120 g d^{-1} and of poultry is 30–40 g d^{-1}. Fish intake has increased by 244% from 1960–1990, from 18–62 $\pm$ 32 g d^{-1}. The main fat in the Mediterranean diet is still olive oil. The mean olive oil consumption is 70–80 g d^{-1}. The legume and pulses intake is still high in the Mediterranean diet. Mean legume and pulses intake is 60–80 g d^{-1}. Vegetable intake has increased whereas fruit intake has decreased over the past 30 years. Mean vegetable intake today in Mediterranean is 280–300 g d^{-1}, whereas fruit intake is 250–300 g d^{-1} [175, 176, 237].

During the nuclear breakdowns (bomb explosions) and nuclear accidents in nuclear stations, great amount of fission products are released in the atmosphere. During the Chernobyl accident radioactive substances had reached a height of 1,000 m (3) and during the first week these cloud-shaped substances had started toward the Scandinavia and from there toward Central Europe and the Balkan countries. According to the existing data, radioactive substances first had appeared in the atmosphere of Northern Greece in the morning hours of May 2, 1986, and in Athens in the afternoon on that same day. The rainfall during the following few days (3–7 May) contributed to the atmospheric purification and transfer of fission products to the ground surface (streets, roads and soil). The total fallout of radioactive dust, measured in Thessaloniki was 150 kBq/m^2.

The approximate values for each of the basic Balkan capitals are ^{131}I (50 kBq/m^2), ^{134}Cs (5 kBq/m^2), ^{132}Te (30 kBq/m^2), ^{140}Bq (17 kBq/m^2), ^{137}Cs (12 kBq/m^2), ^{103}Ru (27 kBq/m^2), ^{106}Ru (6 kBq/m^2), ^{141}Ce (1.9 kBq/m^2), ^{144}Ce (1.6 kBq/m^2), ^{95}Zr (0.5 kBq/m^2), ^{90}Sr (1.4 kBq/m^2). The radiation values in Athens were approximately three times lower. In comparison to them, the total radiation in Muenchen, Germany was almost twice higher (277 kBq/m^2) due to the high concentration of ^{131}I (92 kBq/m^2), and that of ^{132}Te (120 kBq/m^2). It must be noted that the total values for Cs radiation in central Europe as a result of the nuclear weapon tests (1966–1967) were about 4 kBq/m^2.

The exposure of Greek people to natural gamma radiation of terrestrial origin had been an object of concern for investigations in the past [8, 32, 52]. The evaluation of the external doses due to the terrestrial natural sources of ionizing radiation is important, because of their contribution to the collective dose of the population. It is known that over 70 natural radionuclides are present in the surface soil and rocks. Most of them belong to the uranium, thorium and actinium decay series. High concentrations of natural radionuclides, are observed in phosphate soils, igneous rocks and in shales, while low values have been reported for sedimentary rocks. The fission products have mostly been affected in Northern Greece, compared to the other countries of the Balkan region, after the Chernobyl accident.

Regardless the great number of literary data, the national interests, taking in consideration the Chernobyl accident, imposes a thorough study of the world nuclear production, the state of nuclear power stations and their impact on the ecological state of the country.

The results obtained from these studies are used as a regulator of the tendencies for increasing the nuclear energy, working out models for control of the risk of people's exposure as well as the radioactive contamination of the air, soil and vegetation near the source of radiation.

1.4 Formulation of Purposes and Tasks

Natural and artificial atmospheric radioactivity comes from either NPP operation or from accidents in NPPs and they were given special attention after the Chernobyl accident. Due to this, the confrontation of the consequences at the Fukushima accident (in 2011) was more effective, but still, there is not enough knowledge to prevent it. In order to achieve this goal the governments' pressure to their Nuclear Companies to follow in full the IAEA rules for the NPPs safety is of high importance.

Many researches attempted to explain the mechanism of radioactivity's global transfer in the atmosphere. There are enough scientific studies on the diffusion of radioactivity through the air, soil, plants, and foods under certain conditions, mainly for countries which have been affected by the Chernobyl accident. A modeling platform for radioactive transfer processes should be developed on the basis of these studies. The results from such studies should also enable objective forecasts with regard to future environmental radioactive state as well as for undertaking preventive measures against dangerous situations.

The studies on the radioactive contamination transfer in air, soil, plants, foods aim at meeting the requirement for scientific information in that field, under the concrete conditions in Greece, which makes it an actual work for the country.

To achieve this purpose, the following tasks are assigned and presented in this book:

1. Study of airborne radioactivity and radionuclide identification in basic regions of Greece before and after the arrival of the Chernobyl cloud.

2. Study of the diffusion of radionuclides in the environment and theoretical aspects of the diffusion phenomena of radioactive substances in the environment.
3. Study of the impact of atmospheric contamination on soil, plants and foods.
4. Study of the concentration and transfer of ^{137}Cs and ^{40}K in Greece from soil to plants.
5. Study of the mechanism for transfer of ^{131}I and ^{137}Cs in the foodstuffs: dairy, milk products and fresh-water fish.
6. Study of the radioactive decontamination and irradiation process of milk and milk products.
7. Study of the public health impact due to the natural radioactivity of the building materials.
8. Investigations of the marine radioactivity.

Chapter 2
Methods and Materials of Analysis

The research work has been carried out in Greece, mainly in the region of Northern Greece as well as its neighboring geographical regions. The research work aforesaid has been made during the period 1989–1998.

Analyses have been made in the laboratory of Applied Physics at the Alexander Technological Educational Institute of Thessaloniki, which is under the control of the Greek Atomic Energy Committee (GAEC) as well as in the laboratory of the Environmental Radioactivity Measurements, "Demokritos" NRC, Athens.

A great deal of information from data and found correlations, obtained in the world laboratories, has been used in the research work.

- *Object of Research.* Air, soil, plants, food.
- *Methods of Work and Analyses.* Samples are taken and processed according to international accredited methods, ensuring of their state and contents when being studied in the laboratory.

A great number of the airborne radioactivity measurements in Northern Greece for the period 1981–1991 have been obtained. On the basis of the obtained data from studies of the basic physical parameters—temperature, relative humidity, wind speed, atmospheric pressure of the air and rain, a comparison to previous periods as well as anticipations about the future state are made by application of mathematical models.

For estimation of the radioactivity level in a given region, a systematic method has been used, where the distribution is estimated by measurement of the radioactivity on the surface soil Fig. 2.1.

2.1 Methods for Analyses of Atmospheric Radioactivity

Atmospheric radioactivity and the airborne nuclides have been studied by measuring the radioactivity of the dust falling on the earth through filters (IAEA 1989). A constant flow rate air suction pump of AEG, operating at 1.1 m^3h^{-1}, equipped

F. K. Vosniakos, *Radioactivity Transfer in Environment and Food*,
Environmental Science and Engineering, DOI: 10.1007/978-3-642-28741-1_2,

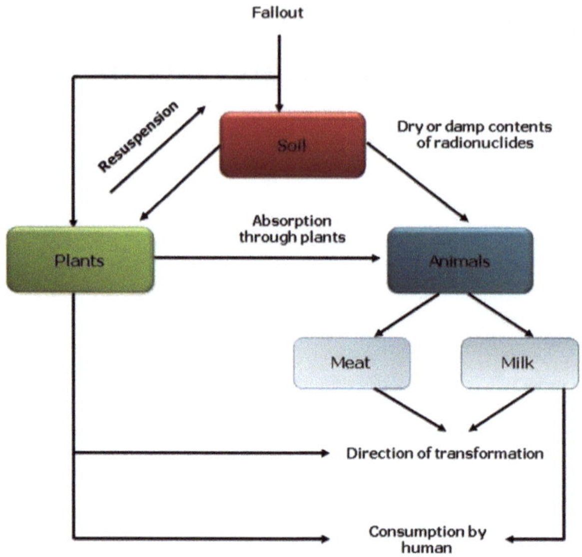

Fig. 2.1 Contamination principal transfer pathways in the environment

with paper filters "Whattman–41", mounted at air inlet position, was used as the radionuclide collection device in Thessaloniki (40.5°N, 22.9°E). The filters are usually compressed to provide a standard counting geometry and are measured by gamma-spectroscopy. The spectroscopy system consisted of a high purity coaxial Germanium detector *p*-type (CP—2100 Tennelec). The sample chamber was a cylinder 12 cm in diameter and 25 cm in height and it was shielded by 5 cm of lead and 0.5 cm of copper in order to minimize the background. The full width at half maximum (FWHM) of the system was found to be 1.95 KeV at 1332 keV of ^{60}Co. The linearity of the detector was checked with a ^{152}Eu source and a simple regression analysis gave a straight line with a correlation coefficient of 0.999 [7, 76, 141, 182].

A radioactivity measuring device was also used and it has been collecting data permanently since 1981 for an industrial area where no nuclear stations exist. The study of the radioactivity transfer is complied with the illustrated device in Fig. 2.2.

2.1.1 Diffuse Radioactivity Measurements

The diffuse radioactivity measurements and the radionuclides identification is made by gamma-ray spectroscopy for identification of ^{103}Ru, ^{106}Ru, ^{134}Cs, ^{137}Cs and ^{131} I presence in the atmosphere 10 days after the arrival of the cloud, in regions endangered by contamination to a greater extent (Sindos—a suburb of Thessaloniki).

Two Geiger-Muller detectors—one for α, β, γ, and the other for β, γ radiation were used with a Leybold Heraeus counter-amplifier system. The detectors were positioned at 3.5 m height from ground. Measurements were made hourly on a 24 h basis and were being stored into an electronic memory unit connected to the

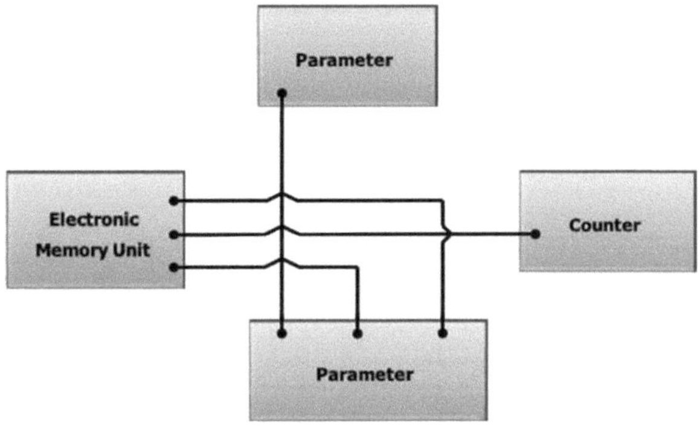

Fig. 2.2 Diffuse radioactivity measuring facility

measuring equipment (Fig. 2.2). Recorded values were in counts per minute (cpm) and were used to estimate the mean daily values for the years 1981–1987.

2.1.2 Radionuclides Detection and Identification

A suction pump of AEG, equipped with paper filters "Whattman–41", mounted at air inlet position, was used as the radionuclide collection device. A time switch circuit was used with the pump to ensure an 8/43 working time over cease time ratio. The suction rate was measured accurately and found equal to 1.1 m^3h^{-1}.

Whattman filters have been analyzed using the γ-ray spectroscopy technique. The associated facility consisted of a Pb-shield, NaI crystal, a photomultiplier tube, a Caberra multichannel analyzer, a preamplifier and a linear amplifying unit, a computer and a printer. Computer printouts contained energy peaks, each peak being associated with a specific radionuclide (Fig. 2.3). The location of each discrete peak on the KeV energy scale, defines the energy of the specific gamma that has created the peak in question. This gamma is characteristic of a particular radionuclide, which is then identified by referring to tables of gamma ray spectroscopy. Finally the total number of events for each peak enables the estimation of the concentration of each radionuclide in air by making use of an empirical relationship as shown in the discussion [190].

2.1.3 Correlation with Meteorological Parameters

A complete meteorological station has also been working since 1981 alongside the radioactivity measuring facility. A data bank of parameters such as ambient temperature, relative humidity, atmospheric pressure and wind were elaborated with the radioactivity data bank to give corresponding mean daily values for the years 1981–1987. A computer program has been used to correlate radioactivity and meteorological parameters on the basis of a linear variation [191].

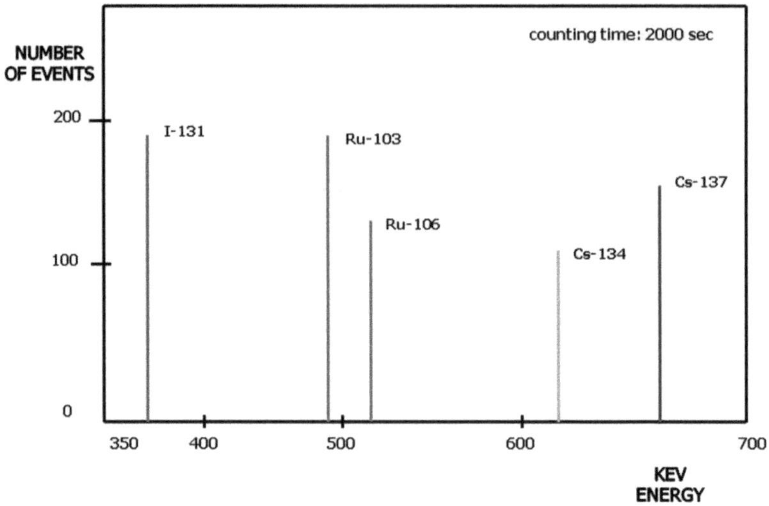

Fig. 2.3 Spectrograph, showing the existence of radionuclides in air

A spectrograph, showing the existence of radionuclides in air has been used (Fig. 2.3).

Anticipation of the environmental contamination level is made on the basis of data of the meteorological conditions and the sources of contamination. On the basis of that transitory and lasting forecasts have been developed.

The total penetration and transfer of the air contaminants is found by simultaneous measurement of the concentration of SO_2, O_3, NOx in the air in the industrial area of Thessaloniki, the meteorological parameters and the concentration C(o) of SO_2 on the ground surface for a definite point of the region [182].

A spectrophotometer BREWER MARK IV has been used for measuring the intensity of solar ultraviolet radiation in regions with wavelength in which the absorbing spectra of SO_2, O_3, and NOx have a special configuration. For calculations of the total amount through the measured intensities, there has been used database of A. W. Brewer [7] and the atmospheric contamination team in Canada [15].

2.2 Methods for Analysis of the Impact of Atmospheric Contamination on the Food Chain (Soil, Plants, Foods)

2.2.1 Analysis of Soils and Plants

Diluting the elements in water and then mixing them with the ground surface study the relative uptake of various radionuclides from soils. Various concentrations of elements which were present in the tested crops were examined before and after soil contamination, as follows: strongly concentrated 10–1,000; slightly concentrated 1–100; non-concentrated 0.1–10; slightly excluded 0.01–1; strongly excluded <0.01 [20, 138].

For determination of the aqueous radioactive contamination, the so-called concentration factor was used, i.e. the ratio between the concentration of a given element in a certain organism and its concentration in water. On the basis of a large informational material, there have been worked out tables for concentration factors (CF) of various classes of fresh water organisms.

Natural radioactivity in the soils in Northern Greece is studied from soil samples, taken from 0 to 20 cm in depth, by special metallic samplers. Samples have been dried, pulverized and sifted through a 2 mm sieve. They were then sealed in small cylindrical pots, made from urea. Samples stayed thus for at least 3 weeks, in order that radioactive equilibrium was obtained between the studied isotopes ^{226}Ru and ^{222}Ru. A highly resolution gamma-spectrometer for measurements was used with a detector HpGe (Tennelec Coaxial Ge Detector System) with relative efficiency of 20% and resolution (FWHM) 1.8 keV of ^{60}Co. The detector was connected with a 4,096-channel analyzer. The measuring time for soil samples was usually 76,000 s. Activities are calculated in Bq kg^{-1}, by using data from the spectrographs. Depositions were calculated in Bq m^{-2} using activities, soil densities and hypothetical depth of 1 cm. The absorbed external dose rates in air were calculated by using the conversion factors [119].

Statistical analysis uses the methods of descriptive statistics as well as these of deductive statistics. And especially the average values, the standard deviations, maximum and minimum values were found for Macedonia and Thrace, as well as for each country separately. Then there were made tests of single groups for estimating the statistical differences of mean levels of activity concentrations.

The level of ^{137}Cs in the soils in Greece one decade after the Chernobyl accident was measured during the period January 1993–May 1995 on 380 soil samples from the soil surface of 5 cm in depth (Fig. 2.4).

All the samples were sealed in containers for at least 15 days in order that equilibrium of isotopes ^{226}Ra and ^{222}Rn was obtained [225, 226].

The taken soil samples after being dried were weighed and measured in Marinelli beakers with gamma-spectroscopy system, consisting of a coaxial Germanium detector of high purity, *p*-type (CP 2100, Tennelec). The sample chamber was a cylinder 12 cm in diameter and 25 cm in height and it was shielded by 5 cm of lead and 0.5 cm of copper, ensuring lowered background. The full width of a half maximum (FWHM) of the system was found 1.95 keV at 1,332 keV of ^{60}Co. The linearity of the detector was checked with ^{152}Eu source and a simple regression analysis gave a straight line with a correlation coefficient of 0.999. The radionuclides used were supplied by "The Nucleus", Oak Ridge, USA. The measuring time was 86,400 s for all soil samples, in order for the time to be sufficient to measure almost any level of radioactive contamination (from 0.1 kBq/m^2).

Firstly, the activity per unit mass is evaluated for each sample. The activity per unit area (Bq/m^2) was calculated afterwards by assuming the soil sample volume

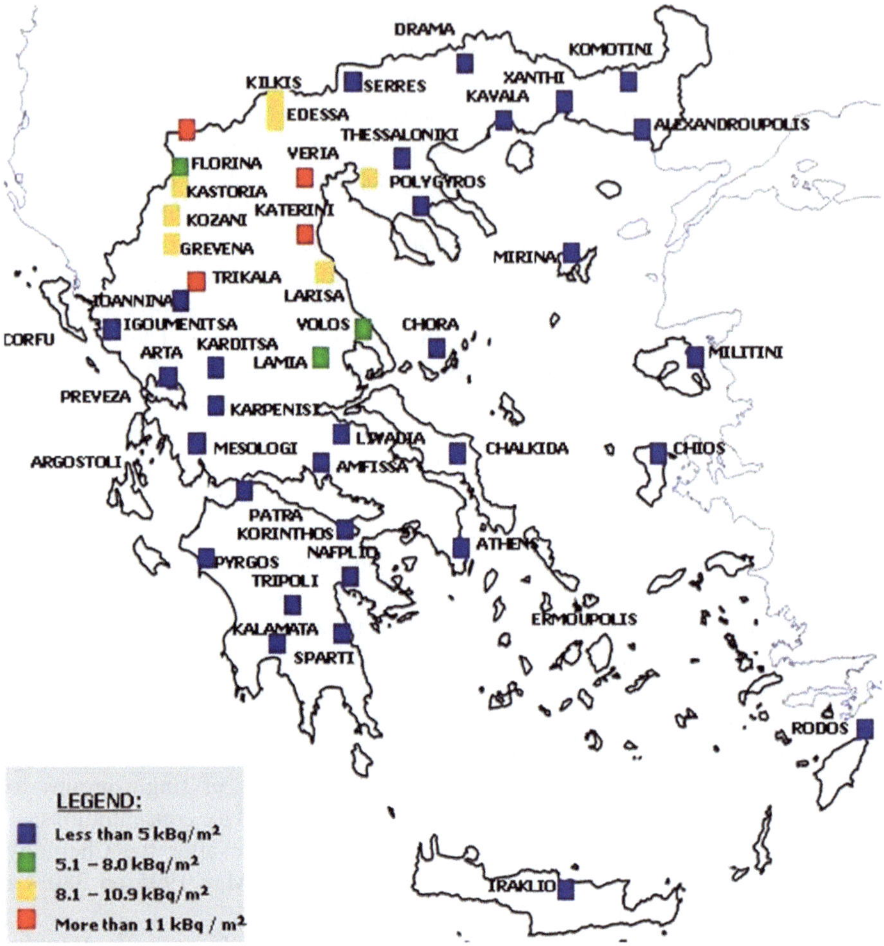

Fig. 2.4 Map of radioactive contamination in Greece

equal to the surface of a layer with depth 1 cm. The mean value and the standard deviation of the density of the 380 samples was 1.43 ± 0.23 Kg/L.

The soil contamination in Greece during the period January 1993 to June 1997 was studied. For that purpose, 742 samples were taken from soil surface (0–5 cm) and from deeper soil layers (5–50 cm), in which the expected mobility of Cs was slight (0.2 y^{-1}). The sampling of surface soil of about 500 cm^3 each, were taken from geographic divisions of Greece with emphasis to those where in 1986 serious depositions of ^{137}Cs (from 15 kBq/m^2 and more) were observed [20, 138, 181].

Samples were studied according to the aforesaid method.

For comparison of the levels of soil contamination according to data during 1986 and 10 years after the Chernobyl accident, the level of ^{40}K (0.0118% of natural potassium in the same soils) was studied [225].

The levels of ^{137}Cs and ^{40}K were compared in the grown plants in 1986 as well as in the cultivated crops during the radioactive fallout. The measuring system was calibrated for the energy using standard sources of ^{137}Cs and ^{60}Co, supplied by "Nucleus", Oak Ridge, USA. The ratio channel-energy was linear, with proportionality coefficient 0.99999.

The activity was calibrated with standards dry soil and dry plants of a known activity (Bq), supplied by the National Research Center in Greece (Demokritos).

In September 1998 samples were taken from corn and clover and soil samples from cultivated land (root depth) in the area of Thessaloniki. Plants were separated from their roots and grain, thoroughly washed, air-dried and oven-dried (80°C). Then they were weighed (dry plant mass) and burned in another oven (Nabertherm) at 500°C for at least 24 h to get carbon white ash. This was done because Cs in a certain volume of ash is 8–9 times more than the Cs in the same volume of dry plant. It should be mentioned that Cs is not lost by oven drying at 500°C (boiling point of Cs and K, 687°C and >750°C). Plant ash was filled in the standard geometry white cup (cylindrical, 7 cm diameter and 2 cm height), recommended for measurement by the Greek National Research Center–Demokritos).

Soil samples were also air-dried, oven-dried (80°C), pulverized, sifted (2 mm sieve) and weighed (mass dry soil). Then they were put to fill the white standard cups. Measuring method: gamma-ray spectroscopy [141].

The concentrations of ^{137}Cs and ^{40}K were measured in kBq/m^2 dry plants. Soil samples were prepared according in the aforesaid manner and their concentrations were measured in kBq/m^2.

For the study of ^{137}Cs and ^{40}K transfer soil-plant in the areas Drama (Eastern Macedonia) and Thessaloniki (Central Macedonia), the corn was taken for its being the main food of cows in the country [10].

Samples were taken in November 1998–1915 plants in total. Each plant was taken out carefully so that it remained intact (roots, plant, grain), as well as the soil round its roots. Plants were gathered when ripe and prepared for analysis according to the manner aforesaid [226].

2.2.2 Foods Analysis

Studies of radioactive contamination of foods have been made: of cow milk, milk products, cheese, curd and yogurt and also typical representatives of freshwater fish [186, 187, 188, 189].

The determination of the radioactivity of cow milk to cheese curd, cheese whey, mytzithra, and lactose serum was made by application of gamma-spectroscopy. The spectroscopic system consisted of:

- *Detector*. NaI (Tl) scintilator $2'' \times 2''$.
- *Shield*. 50 mm Pb + 5 mm Cu cylinder.
- *Analyzer*. Canberra 8100 (USA), multi-channel analyzer with 4,096 channels.

- *Resolution (FWHM).* 70 keV approximately for ^{137}Cs or 10.5%.
- *Geometry of the chamber.* cylindrical, dimensions 120 × 250 mm.

The efficiency of NaI (Tl) was determined by a set of gamma-ray standard sources (^{152}Eu, ^{134}Cs, ^{137}Cs), supplied by Amersham. For the energy calibration the following radionuclides were used:

^{131}I	364.5 keV
^{137}Cs	661.6 keV
^{134}Cs	604.7, 795.8 keV
^{152}Eu	122, 245, 344, 427.5, 779, 964, 1,099, 1,408 keV

The radionuclides used were supplied by "The Nucleus", Oak Ridge, USA.

Fats determination in the milk, the cheese whey and the lactose serum were done according to Gerber's method (standard 1/1963 IDF). The Shmid Bondzinski process (standard 3/1963 IDF) was used for the curd and the mytzithra. The Kjeldahl method (standard 20/1962 IDF) [86, 97] was applied for determination of the total amount of protein and non-protein nitrogen (N. P. N.). The proteins content was calculated by using a nitrogen conversion factor of 6.38. The spectroscopic method was used for the determination of lactose [113, 136]. The ashes as well as the total solids in the milk, in the cheese whey and in the lactose serum were determined according to standard methods (British standard 1741/1963, standard 21/1962 IDF). In the case of cheese curd and mizithra, the ash and the total solids and NaCl were determined by standard methods (standard 27/1964 IDF, 4/1963 IDF, standard 17/1961 IDF).

The fresh chilled raw milk was supplied by the experimental farm of the Technological Institution of Thessaloniki. The method used for the production of "Feta" cheese was done according to Kosikowski's procedure [104]. Yogurt from the market was used as a starter and the curdling was accomplished by adding artificial rennet (Rennilase 150 L, Novo, Denmark). Mytzithra was traditionally produced according to the following procedure: A mixture from cheese whey and milk, in a 9:1 ratio is heated for 45 min at 90°C, pH being fixed to 5.4 by adding 10% solution of citric acid. The obtained curd is heated for about 20 min until it reaches 90°C and then it is pressed for 16 h at room temperature.

Two experiments have been carried out for determining ^{131}I transfer from "Feta cheese" to clear water or brine and conversely.

The first experiment includes "Feta cheese", which had been made in the experimental laboratory of the Alexander Technological Educational Institution of Thessaloniki from contaminated with ^{131}I cow milk (10–15 kBq/kg).

Portion of the cheese, 6 × 6 × 6 cm sized were divided into 4 zones and immersed in different solutions of NaCl, KCl and in tap water. The immersion time varies from 5 to 48 h. According to the immersion, there have been outlined four zones of the cheese, 0.75 cm thick from external to inner surface (see Fig. 2.5). These four zones have been separately removed and homogenized in a blender and the radioactivity was measured into 50 g samples for 500 s.

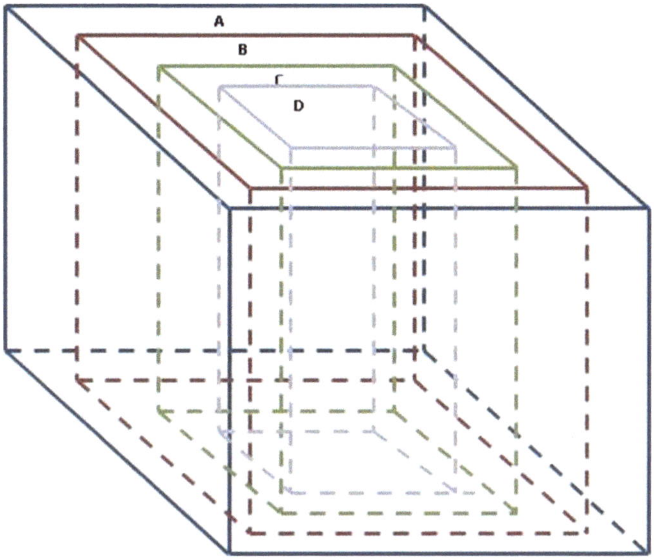

Fig. 2.5 Section of the cheese layers

The second experiment was carried out with "Feta" cheese, which had been bought from the market. The cheese pieces used were of the same size as in the first experiment. Three zones, 1 cm thick were formed after immersing the cheese in solutions of NaCl, KCl and tap water, previously contaminated with ^{131}I (10–12 kBq kg^{-1}). The cheese samples were examined with respect to their radioactivity as in the first experiment.

^{131}I impact on the lactic acid microflora of the yogurt was examined in two varieties of yogurts: the set yogurt and the strained yogurt. Milk was standardized with respect to its fats content (3.9% for cow milk and 6.00% for sheep milk), it had been boiled for 5–10 min, cooled to 45°C and inoculated with 2% starter culture from Thermophilus and bulgaricus [37, 38, 151]. Milk had been inoculated for 3 h at 45°C temperature. The fermented milk had been consequently stored in refrigerator at 4–5°C for 12 h. The yogurt variety was obtained by straining the tissue matter, upon which a part of the serum was retransferred [16, 37, 38, 151].

For decontamination of milk from radionuclides (^{131}I, ^{134}Cs, ^{137}Cs), ion-exchange resins for the unit chosen for this purpose were used. The elements of this decontamination unit has been designed according to the calculations for a capacity of about 5,000 l milk per day at constant flow [197].

The selection of the resin was made taking into account:

- The avoidance of the entrance of harmful substances in milk
- The avoidance of organoleptic changes, and
- The capability of ion-retainment.

Amberlite IRA 120 Plus and Amberlite IRA 904 as cationic and anionic exchangers were used respectively. Their grain size of the resin was chosen so that to allow the liquid pass freely through the columns, without causing changes in the pressure, which had not been foreseen upon designing the unit. After running out, the resin should be regenerated according to the manufacturer's instructions. Regeneration of the cationic resins was obtained through their passing in a diluted solution of HCl (about 6%), and the anionic resins with respective solution of NaCl.

A fixed bed anion-cation removal method has been used for the contamination, which uses columns with ion-exchanger for the accomplishment of the final purpose.

For the laboratory tests glass columns were used with the respective ion-exchangers, as well as cow milk with the following chemical analysis:

Total acidity	17.00°D
Proteins	3.06%
pH	6.62
Fats	3.56%
Lactose	5.52%

Similarly, for long-life skim milk which was also used, the characteristics were: total acidity: 15.5 (°D); fats: 1.5%; proteins: 3.5%.

The experimental procedure followed the steps:

- Adding solutions of radioactive iodine (^{131}I) and radioactive cesium (^{137}Cs) in ionic form in the milk. The concentration of the radioactive substances added was about 1,000 times higher than it was expected to be in order to control the effectiveness of the method. The manufacturer of ^{131}I was N.R.C. Demokritos (Greece) and the one of ^{137}Cs—the Nucleus Company (USA).
- Heating to about 30°C.
- Passing through the column of the anionic resin (^{131}I removal).
- Acidifying with a diluted solution of citric acid to pH 5.3. The avoidance of excess of this value is very important in order to avoid the formation of colloids, which impede the normal flow of the milk through the column, but also deterioration of the milk, as it has been observed during the experimental work.
- Passing of a cationic ion-exchanger through the column (removal of $^{134,\ 137}$Cs).
- Neutralization of milk by diluted solution of KOH (pH 6.5)
- Radioactivity control
- Quality control of milk.

Comparative studies were carried out in two cultured fresh-water fish, *Cyprinus carpio* (*carp*) and *Anguilla anguilla* (*eel*), to determine their tolerance in the uptake of ^{137}Cs (3,000 Bq/l) [280].

The histological studies were concentrated in muscular tissues, livers, kidneys and gills. The symptoms observed include hypothermia, and gradual degeneration of liver and kidney tissues. The physiology and anatomy of each species played an important role in the accumulation process of ^{137}Cs.

The determination of ^{137}Cs was made by γ-spectroscopic system, which was described previously. Fish were cultured in small water tanks, artificially contaminated with radioactive ^{137}Cs. The species *Anguilla anguilla* were collected from artificial ponds 2 days before the experiment. The fish *Cyprinus carpio* were collected from a local lake. They were placed in a 200 l tank, supplied with good aeration and a continuous through flow of tap water, which had been dechlorinated by active carbon. The dimensions of the water tanks used were 79 cm in length, 35 cm in width and 50 cm in height.

The fish were sacrificed everyone or 2 weeks, weighed, their length was measured and their overall conditions were compared to the control. These observations were recorded and the radioactivity of a few organs was measured. The samples of the organs were fixed with 10% formaldehyde solution, embedded in paraffin wax and sections of 5–10 μm thickness stained with Erlich's hematoxylin-eosin, to be examined under the microscope.

The results were processed by modern computer technologies. On the ground of that, condition models were obtained for forecasting the changes in radioactive contamination and the radioactivity transfer through the chain air-soil-plants-foods. The results and the found correlations are presented in tables and figures.

Chapter 3
Results and Discussions

3.1 Studies on the Atmospheric and Environmental Radioactivity

3.1.1 Monitoring Airborne Radioactivity in the Industrial Area of Thessaloniki

The measuring method for determining the radioactivity in the air is based on an air-filtering method where the atmospheric air is pumped out and analyzed by γ-spectroscopy, in order to measure the presence of radioactive isotopes [32].

It has been found that ^{214}Pb, ^{133}Ba, ^{85}Sr or ^{85}Kr, ^{103} Ru and ^{40}K are the most frequent occurring isotopes in the examined filters with pumped in atmospheric air [182].

The results are presented in Table 3.1. The radionuclide variation (focused on the five radionuclides) for 1989 and 1999 is illustrated in Fig. 3.1. A generally observed fact is the increased presence of ^{40}K in comparison with the other nuclides that has been the basic variation since January 1989 to June 1989 and also since April 1990 to December 1990. In the intervals between these periods, ^{40}K has a constant concentration in the air, amounting to 200 μBq/m^3. The rest traced radionuclides, ^{214}Pb, ^{133}Ba, ^{85}Sr or ^{85}Kr and ^{103}Ru have shown similar behavior for the period January 1989–December 1990. During the summer months of the year 1989, increased radioactivity has been found. It was due to these radioisotopes. Since September 1990, a slight, almost linear, increase of the atmospheric radioactivity has also been observed, except for ^{85}Sr in March 1990 and ^{214}Pb in October 1990. It is obvious that the parameters, which determine the atmospheric radioactivity, are elaborate and it is difficult to be examined separately. Figures 3.2, 3.3 and 3.4 show the annual variation of the meteorological parameters—air temperature, relative humidity and airflow speed respectively during the period 1989–1990.

F. K. Vosniakos, *Radioactivity Transfer in Environment and Food*,
Environmental Science and Engineering, DOI: 10.1007/978-3-642-28741-1_3,

Table 3.1 The most frequent occurring radionuclides in the air

γ-ray Energy(keV)	Isotope	Half-life	Minimum concentration (mBq/m^3)	Maximum concentration (mBq/m^3)
242	^{214}Pb	26.8 m	25	475
355	^{133} Ba	10.7 y	45	595
	^{85}Sr	64.73d		
514	or		55	650
	^{85}Kr	10.73y		
612	^{103}Ru	39.3 d	25	475
1460	^{40}K	1.2710^9y	110	1450

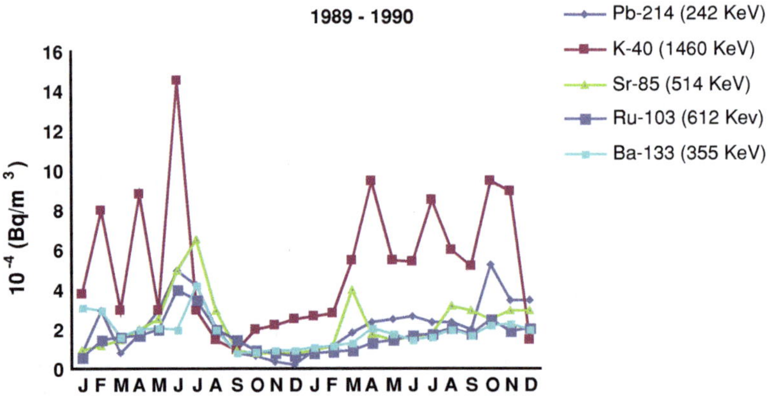

Fig. 3.1 Variation of the airborne radioactivity 1989–1990

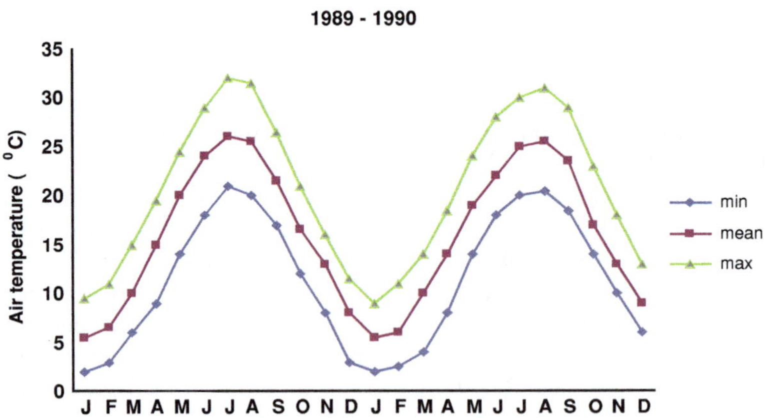

Fig. 3.2 Variation of the air temperature for 1989–1990

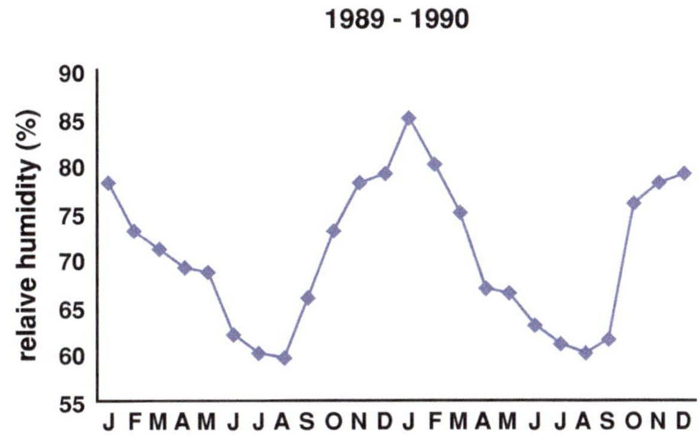

Fig. 3.3 Variation of the relative humidity for 1989–1990

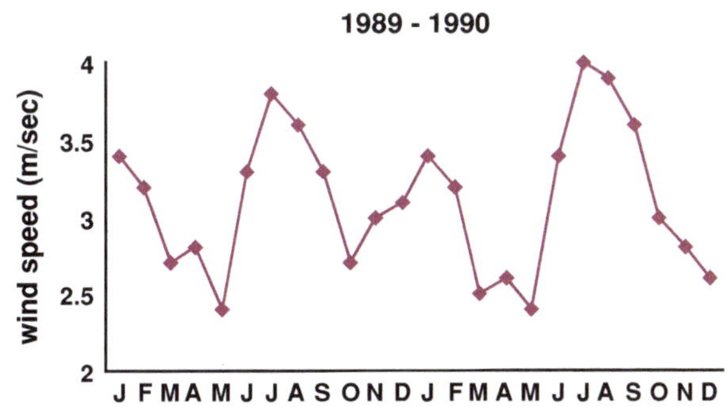

Fig. 3.4 Variation of the wind speed for 1989–1990

The possible relationship between the measured atmospheric radioactivity and the three mentioned parameters cannot be presented by a simple correlation [8]. Besides, such relationships can result by complicated mathematical calculations with the use of different parameters, which are not possible to be determined experimentally easily. It was found empirically, however, that the correlation between the atmospheric radioactivity and the air-speed, humidity or air-temperature, is given by a second-order equation of the following type:

$$y = Ax^2 + Bx + C$$

where: y = radioactivity, x = temperature or humidity or air-speed, and A, B, C are empirical constants which can be determined by experimental data.

The thorough analysis of the above experimental data for the years 1989 and 1990 showed that the relative coefficients (R) for all the isotopes had satisfactory values by which, the above equation describes the interaction between the atmospheric radioactivity and the three meteorological parameters (speed–humidity–temperature). Some typical parameters of this analysis are shown in Fig. 3.5. Figure 3.5 presents the variation between the radioactivity (^{85}Sr) and the air speed according to above equation in 1989, (where A = 102.805, B = −66.491, C = 10.848 and

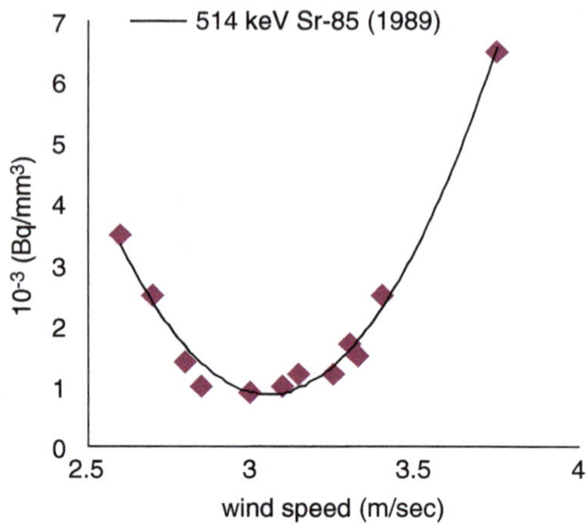

Fig. 3.5 Variation of the airborne radioactivity (^{85}Sr) for 1989 with the wind speed

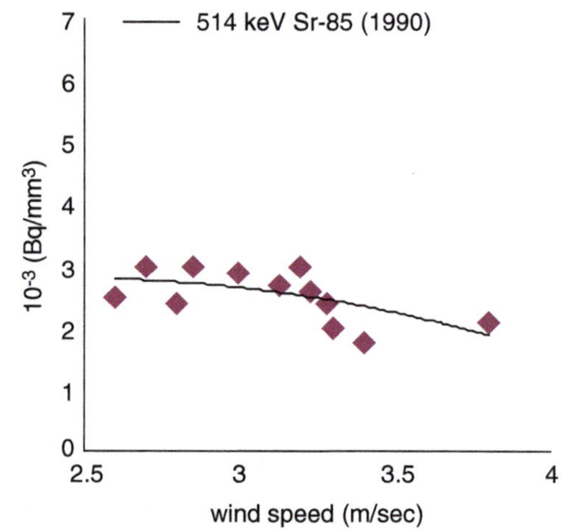

Fig. 3.6 Variation of the airborne radioactivity (^{85}Sr) for 1990 with the wind speed

correlation coefficient, $R = 0.925$). Figure 3.6 presents the same variation for 1990 (where $A = 8.429, B = -2.873, C = 0.287$ and $R = 0.448$). Finally, Figs. 3.7, 3.8 present the variation between radioactivity (^{40}K) and air-temperature according to Eq. (3.5) for 1989–1990 (where $A = -5.047$, $B = 1.375$, $C = -0.037$ and $R = 0.723$) (Figs. 3.9, 3.10).

Also other experiments were carried out: (I)—for study and explanation of the repetition of the empirical correlation between radioactivity and the meteorological parameters (speed, humidity and air temperature) (II)—for finding additional parameters, influencing the atmospheric radioactivity in a definite way, and (III)—for finding the possibility of correlating the radioactive and non-radioactive (SO_2, O_3, NO_x) environmental pollutants.

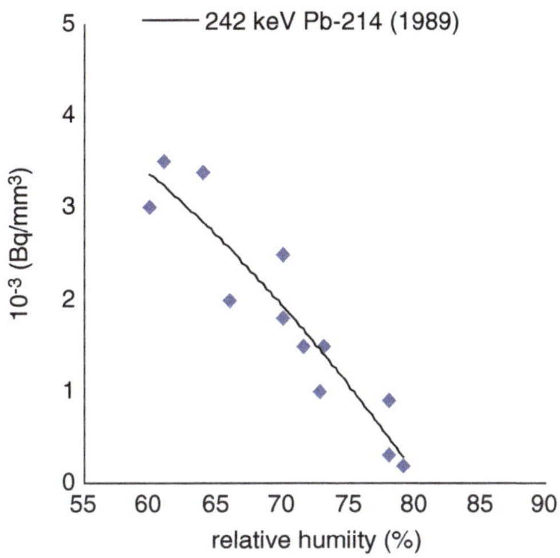

Fig. 3.7 Variation of the airborne radioactivity (^{214}Pb) for 1989 and the relative humidity

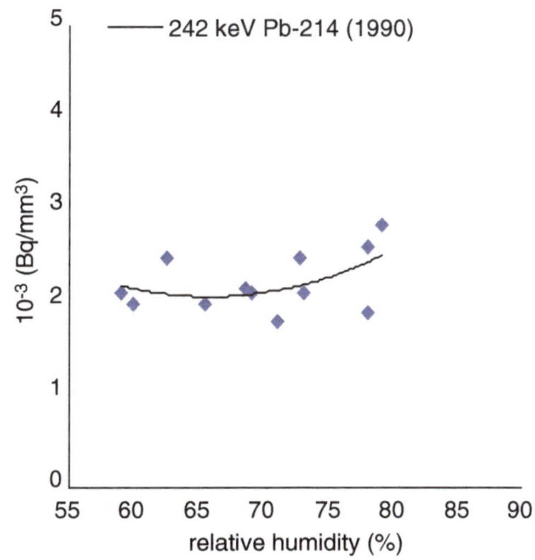

Fig. 3.8 Variation of the airborne radioactivity (^{214}Pb) for 1990 and the relative humidity

3.1.2 Identification of Airborne Radionuclides in the Environment of Northern Greece

The results of the measurements made for the airborne radioactivity in Northern Greece during the three years (1989, 1990, 1991) were used for comparison with the former periods and for developing proper mathematical models determining the temperature and humidity of the air respectively.

The data from γ-spectroscopic analysis of "Whattman-41" paper filters showed considerable concentrations of radionuclides of ^{214}Pb, ^{85}Kr, ^{208}TI, ^{214}Bi, ^{228} Ac and ^{40}K in the atmosphere of Sindos (the industrial area of Thessaloniki). The mean values

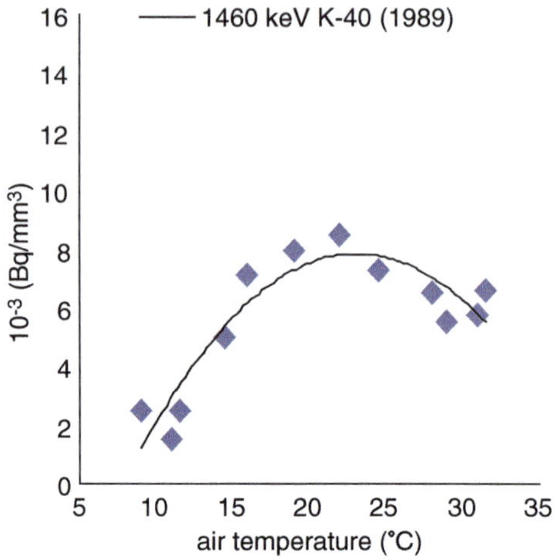

Fig. 3.9 Variation of the air radioactivity (^{40}K) for 1989 with the air temperature

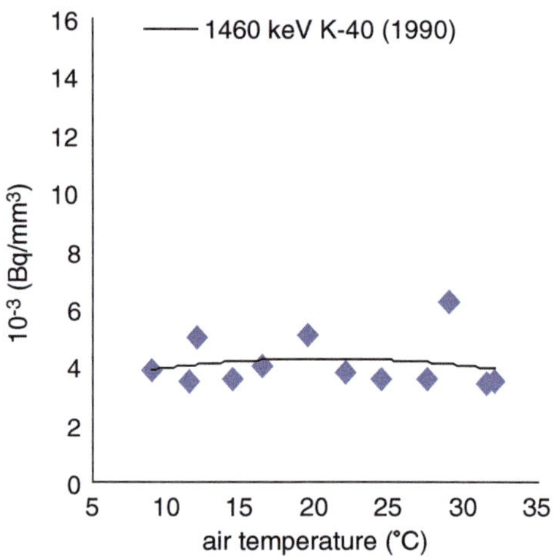

Fig. 3.10 Variation of the air radioactivity (^{40}K) for 1990 with the air temperature

of minimum and maximum concentration of the most observed radionuclides during 1989, 1990 and 1991 are illustrated in Table 3.2 and Figs. 3.11, 3.12, 3.13 and 3.14.

Tables 3.2, 3.3, 3.4 and 3.5 present the air radioactivity for 1989, 1990 and 1991, while Table 3.6 present the meteorological parameters for the same period.

It is obvious that the maximum concentration in the air is that of ^{40}K in comparison with the other nuclides. The rest of the traced nuclides ^{214}Pb, ^{85}Kr, ^{208}TI, ^{214}Bi and ^{228}Ac have similar behavior during the period January 1989–December 1991. During the summer and autumn months there was observed considerate increase in the radioactivity, due to these nuclides. It is also obvious that the parameters, which have influenced the atmospheric radioactivity, are elaborate and it has been difficult to be examined independently. Table 3.6 shows the mean values

Table 3.2 The most frequently observed isotopes in the atmospheric air

γ-ray energy (keV)	Nuclide	Decay chain[b]	Half-life	Minimum maximum concent[a] (mBq/m^3)	
242	^{214}Pb	^{238}U	26.8 m	0.38	3.30
295	^{214}Pb	^{238}U	26.8 m	0.19	2.10
352	Ann.	^{238}U	26.8 m	0.33	2.97
514	^{85}Kr		10.73y	0.35	3.23
583	^{208}Tl	^{232}Th	3.1 m	0.24	2.13
609	^{214}Bi	^{238}U	19.9 m	0.32	2.63
911	^{228}Ac	^{232}Th	6.13 h	0.19	1.60
1460	^{40}K		1.27×10^9y	0.65	7.27

[a] Mean values of the measurements for the three years
[b] The decay chain, in which the nuclide appears

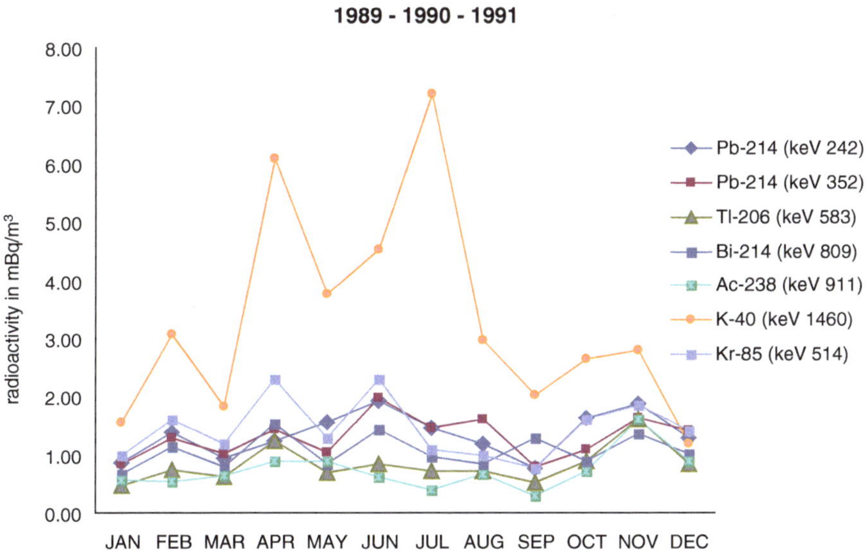

Fig. 3.11 Mean value of the air born radioactivity for 1989, 1990, 1991

of the meteorological parameters—air temperature, relative humidity, air flow speed, atmospheric pressure and rainfalls, during the years 1989, 1990 and 1991.

It is impossible to determine the possible relationship between the measured atmospheric radioactivity and the above-mentioned three meteorological parameters by a simple correlation. In addition, a similar equation is a result of elaborate mathematical calculations by various parameters, which cannot be easily determined through experiments. [8, 52, 180, 190, 191].

An expanded analysis of this experimental data for the years 1989, 1990 and 1991, showed that the correlation coefficients (R) for all isotopes had satisfactory

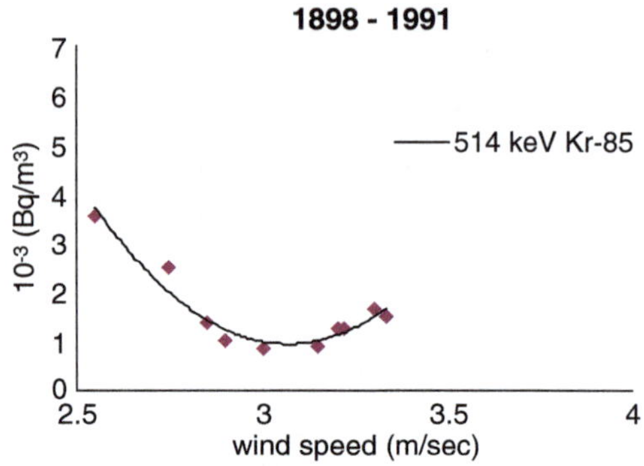

Fig. 3.12 Variation of the mean values of radioactivity with the wind speed during the period 1989–1991

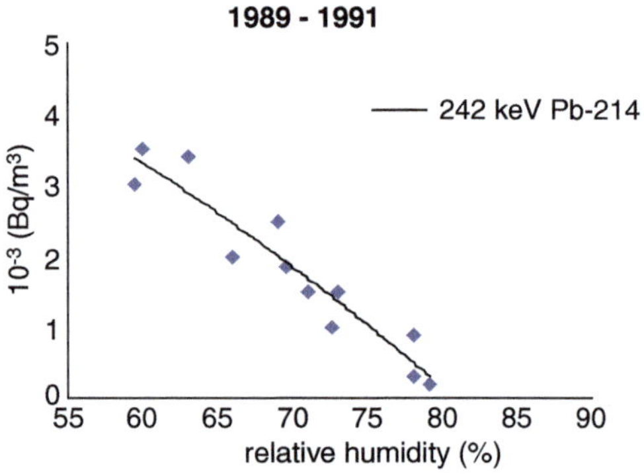

Fig. 3.13 Variation of the mean values of airborne radioactivity for 1989–1991 with the relative humidity

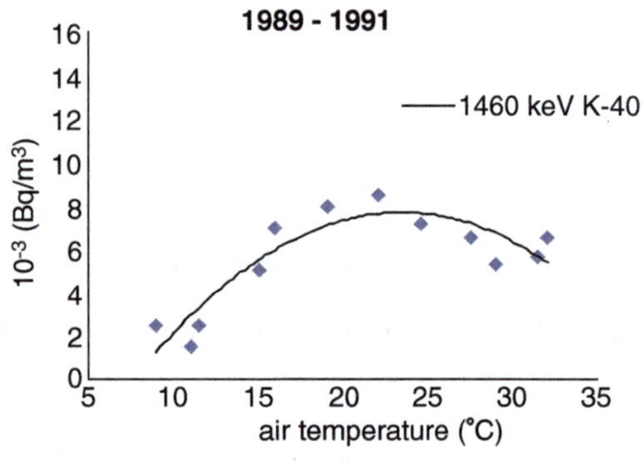

Fig. 3.14 Variation of the mean values of airborne radioactivity with the air temperature

values by the following equation describes the interaction between the atmospheric radioactivity and the three parameters (speed–humidity-temperature). Some typical examples from this analysis are shown in Figs. 3.12, 3.13 and 3.14. Figure 3.12 illustrates the variation between radioactivity (^{85}Kr) and the air speed according to the following equation:

$$y = Ax^2 + Bx + C$$

Table 3.3 Year 1989

Isotope	^{214}Pb	^{214}Pb	^{214}Pb	Ann.	^{206}Tl	^{214}Bi	^{228}Ac	^{40}K
Eγ (keV)	241.9	295.2	351.9	513.9	583.1	609.3	911.0	1460.7
Jan	0.57		0.76	0.95	0.46	0.39		2.50
Feb	1.90		1.90	2.30	1.40	1.30		5.20
Mar	0.57		1.10	0.87	0.70	0.50		1.80
Apr	1.20		1.40	2.50	0.87	1.30		5.60
May	1.70		0.63	0.59	0.30	0.48		1.60
Jun	3.10		4.00	4.30	1.30	3.30		9.60
Jul	2.80		3.10	1.40	1.10	0.90		15.00
Aug	1.30		2.70	0.90	0.70	0.40		1.20
Sep	0.48		0.37	0.27	0.36	0.46		1.00
Oct	0.29		0.49	0.37	0.31	0.26		0.70
Nov	0.27		0.51	0.35	0.30	0.25		1.30
Dec	0.27		0.53	0.32	0.32	0.35		1.50

Table 3.4 Year 1990

Isotope	^{214}Pb	^{214}Pb	^{214}Pb	Ann.	^{208}Tl	^{214}Bi	^{228}Ac	^{40}K
Eγ (keV)	241.9	295.2	351.9	513.9	583.9	609.3	911.0	1460.7
Jan	0.61		0.30	0.30	0.35	0.41	0.28	1.50
Feb	0.67		0.35	0.47	0.32	0.72	0.30	1.60
Mar	1.10		0.57	0.95	0.41	0.76	0.37	3.30
Apr	1.20		1.60	2.40	2.10	1.70	0.90	6.00
May	1.50		1.20	1.60	0.93	0.73	0.80	3.60
Jun	1.60		0.87	1.30	1.10	0.73	0.13	3.50
Jul	1.30		0.93	1.60	0.87	1.60	0.51	5.50
Aug	1.40		1.30	0.96	0.93	1.20	0.68	3.80
Sep	1.10		0.93	1.10	0.63	2.50	0.10	3.20
Oct	3.30		1.50	1.80	1.50	1.60	1.00	6.10
Nov	1.90		1.50	1.90	1.60	1.70	1.30	5.80
Dec	1.90		2.00	2.10	1.20	1.50	0.91	0.77

for 1989, 1990 and 1991 (where A = 102.805, B = −66.491, C = 10.848 and R = 0.925); x = air speed. Figure 3.13 presents the variation between the radioactivity (^{214}Pb) and the air humidity for 1989, 1990 and 1991 (where A = 5.923, B = 0.037, C = −0.001 and R = 0.682); x = relative air-humidity. Figure 3.14 presents the variation between the mean values of radioactivity (^{40}K) and the air temperature, according to Eq. (3.5) for 1989, 1990 and 1991 (where A = 3.226, B = −0.364, C = 0.015 and R = 0.617); x = air-temperature.

Conclusions

1. There was found considerable concentration of the radionuclides ^{214}Pb, ^{85}Kr, ^{208}TI, ^{214}Bi, ^{228}Ac and ^{40}K in the atmosphere of the industrial area of Sindos, Thessaloniki, during the years 1989, 1990 and 1991. It was demonstrated that ^{40}K concentration was comparatively high.

Table 3.5 Year 1991

Isotope	^{214}Pb	^{214}Pb	^{214}Pb	Ann.	^{208}Tl	^{214}Bi	^{228}Ac	^{40}K
Eγ (keV)	241.9	295.2	351.9	513.9	583.9	609.3	911.0	1460.7
Jan	1.40	0.73	1.50	1.70	0.63	1.20	0.87	0.65
Feb	1.60	1.10	1.60	2.00	0.51	1.40	0.80	2.40
Mar	1.20	1.20	1.20	1.70	0.73	1.30	0.93	0.39
Apr	1.30	1.10	1.30	1.90	0.73	1.60	0.87	6.70
May	1.50	0.67	1.30	1.70	0.87	1.30	1.00	6.10
Jun	1.10	0.55	1.10	1.30	0.10	0.30	1.10	0.49
Jul	0.27	0.19	0.33	0.47	0.22	0.41	0.28	1.10
Aug	0.87	0.54	0.87	1.30	0.51	0.93	0.67	3.90
Sep	0.73	0.45	1.10	1.10	0.55	0.87	0.48	1.90
Oct	1.30	0.73	1.30	1.60	0.90	0.80	0.43	1.20
Nov	3.50	2.10	2.90	3.00	2.90	2.10	1.90	1.30
Dec	1.70	1.10	1.70	2.10	1.00	1.20	0.87	1.30

Table 3.6 Mean values of air-temperature, relative humidity, atm. pressure, rain and wind speed for the years 1989, 1990 and 1991

Month	Temperature (°C)	Humidity (%)	Wind speed[a] (ms^{-1})	Atm press (mmHg)	Rain (mm)
Jan	5	80	3.4	786.7	47
Feb	7	77	3.2	768.3	39
Mar	10	72	2.7	768.5	42
Apr	15	67	2.8	759.2	52
May	20	68	2.4	763.1	54
Jun	24	62	3.3	760	43
Jul	26	60	3.8	760.3	44
Aug	26	58	3.6	762.1	30
Sep	23	66	3.3	765	42
Oct	18	70	2.7	760	58
Nov	12	78	2.9	765.7	63
Dec	7	80	3.1	768	60

[a] The wind directions for that period were: Northern 15%, North-Eastern 6%, Eastern 5%, South-Eastern 4%, Southern 6%, South-Western 7%, Western 3%, North-Western 5% and Calm 49%

Table 3.7 Mean values of the air temperature, relative humidity, atmospheric pressure, rainfall and air speed in Northern Greece (1989, 1990, 1991, 1992 and 1993)

Parameter	Monthly mean value											
	JAN	FEB	MAR	APR	MAY	JUN	JUL	AUG	SEP	OCT	NOV	DEC
Temp (°C)	6	8	10	16	21	24	27	28	22	19	11	9
Hum (%)	82	79	74	65	70	64	62	56	68	70	76	83
Wind sp (ms^{-1})	3.5	3.4	3.0	3.1	2.7	3.7	3.9	3.3	3.2	2.8	3.0	3.1
Atm.pr (mmHg)	785	767	767	760	762	761	762	764	766	763	767	769
Rain (mm)	49	40	43	50	56	42	46	34	44	60	62	60

2. An equation was constituted for determining the correlation between the atmospheric radioactivity and the air parameters—speed, relative humidity and temperature.

The empirical constants of the equation $y = Ax^2 + Bx + C$, are determined experimentally with a high reliability.

3.1.3 Measurements of the Atmospheric Radioactivity in Northern Greece

Here is presented the data from daily measurements of air radioactivity in the area of Northern Greece by gamma-spectroscopy for the years 1989–1993. The daily variations of the air masses were also observed for the same years and place.

The mean annual concentrations (mBq m^{-3}) of the most frequently occurring radionuclides for 1989, 1990, 1991, 1992 and 1993, are presented in Table 3.2 and the respective correlation between ^{214}Bi–^{214}Pb, ^{208}Tl–^{228}Ac and ^{40}K–^{137}Cs is illustrated in Fig. 3.11. As one would expect, the correlation between ^{214}Bi–^{214}Pb and ^{208}Ti–^{228}Ac is of great interest as they are daughters of ^{238}U and ^{232}Th respectively. The correlation between ^{137}Cs and ^{40}K reveals that the areas with higher concentration of ^{40}K are less inclined to contamination with ^{137}Cs and conversely [52]. Table 3.3 presents the mean values of minimum and maximum concentrations (mBq/m^3) of the isotopes aforesaid for that five-year period (1989–1993).

It was found by observations that ^{214}Bi, ^{214}Pb, ^{208}TI, ^{228}As and ^{40}K are the most frequently occurring natural isotopes in the examined filters of the pumped atmospheric air, as well as ^{137}Cs, accumulated in the soil after the last accident at Chernobyl [44, 52, 180, 190, 191].

It is clear that the parameters influencing the atmospheric radioactivity are elaborate and it is difficult to be studied separately.

Table 3.7 illustrates the mean values of the meteorological parameters temperature, relative humidity, airflow speed, atmospheric pressure and rainfalls in the major area of Northern Greece for 1989, 1990, 1991, 1992 and 1993.

Taking into consideration the above-mentioned relationship between the measured atmospheric radioactivity and the previously mentioned meteorological parameters, it is difficult to be given by a simple correlation. In addition, a similar correlation can be obtained through elaborate mathematical calculations by using various parameters, which cannot be determined experimentally.

It was found experimentally that the correlation between the atmospheric radioactivity and the air parameters—speed, humidity and temperature, is given by a second-order equation [190].

On the other hand, the atmospheric fallout from nuclear stations is diffused in several ways and so it contributes to a greater radioactivity.

The radioactive substances that are diffused in the atmosphere are taken by the air or by the soil and turned around. The higher the speed is the more effective is

Table 3.8 Most frequently occurring radionuclides in the air in the years 1989, 1990, 1991 and 1992

Isotope	Airborne radioactivity (mBq m^{-3})			
	1989	1990	1991	1992
^{214}Pb	18.06 ± 4.12	21.97 ± 4.83	20.59 ± 2.48	13.66 ± 1.82
^{208}Tl	10.14 ± 1.94	14.93 ± 2.16	12.07 ± 1.41	8.65 ± 1.50
^{214}Bi	12.36 ± 1.85	18.95 ± 2.21	16.76 ± 2.77	12.79 ± 0.87
^{137}Cs	14.85 ± 4.31[a]	n.d.[b]	4.34 ± 0.41[c]	15.97 ± 2.02[d]
^{228}Ac	n.d.	4.84 ± 0.82	12.74 ± 1.50	6.58 ± 0.75
^{40}K	58.75 ± 4.06	55.83 ± 2.23	34.25 ± 1.54	37.85 ± 1.48

[a] Mean value for three months. During the rest of the months, concentration was unnoticeable
[b] Unnoticeable
[c] Mean value for six months
[d] Mean value for 9 months

the dilution of gaseous and solid radioactive substances in the atmospheric air. The smallest substances quickly reach the highest atmospheric layers and therefore these do not influence the ground. The rest substances reach near the ground layers and their deposition in the soil causes decrease in the concentration of radioactive substances in the air and conversely (dry deposition) [53, 191].

Rain also contributes to the decrease of the atmospheric radioactivity. In this case the radionuclides are either "crumbled" (rained out) or "rinsed" (washed out). Both the procedures constitute the process of the "fallout". All the rest of the radioactive substances, which still remain in the atmosphere are continuously dissolved until they cannot be measured and become harmless to the environment.

Conclusion

The data has been obtained by γ-spectroscopic analysis with "Whattman-41" filters show the concentrations of ^{214}Pb, ^{208}TI, ^{214}Bi, ^{137}Cs, ^{228}Ac and ^{40}K in the atmosphere in the major area of Thessaloniki [53]. The mean annual concentrations (mBq m^{-3}) of the most frequently occurring radionuclides 1989, 1990, 1991, 1992 are presented in Table 3.8 and the mean seasonal concentrations are illustrated in Fig. 3.15a, b, c.

Table 3.9 presents the mean values of minimum and maximum concentrations (mBq m^{-3}) of the isotopes aforesaid for the period of four years (1989–1992). It is obvious that ^{40}K has maximum concentration in the air and increased presence in the environment, in comparison to the rest isotopes. ^{214}Pb, ^{208}TI, ^{214}Bi, ^{137}Cs, ^{228}Ac show similar behavior for the period January 1989–December 1992. During the summer and the autumn months, a considerable increase of radioactivity was observed due to these isotopes. It is obvious that the parameters influencing the atmospheric radioactivity are elaborate and it is difficult to be examined separately.

Table 3.10 illustrates the mean values of the meteorological parameters air temperature, relative humidity, wind speed, atmospheric pressure and rainfall in the major area of Thessaloniki for the years 1989, 1990, 1991, and 1992.

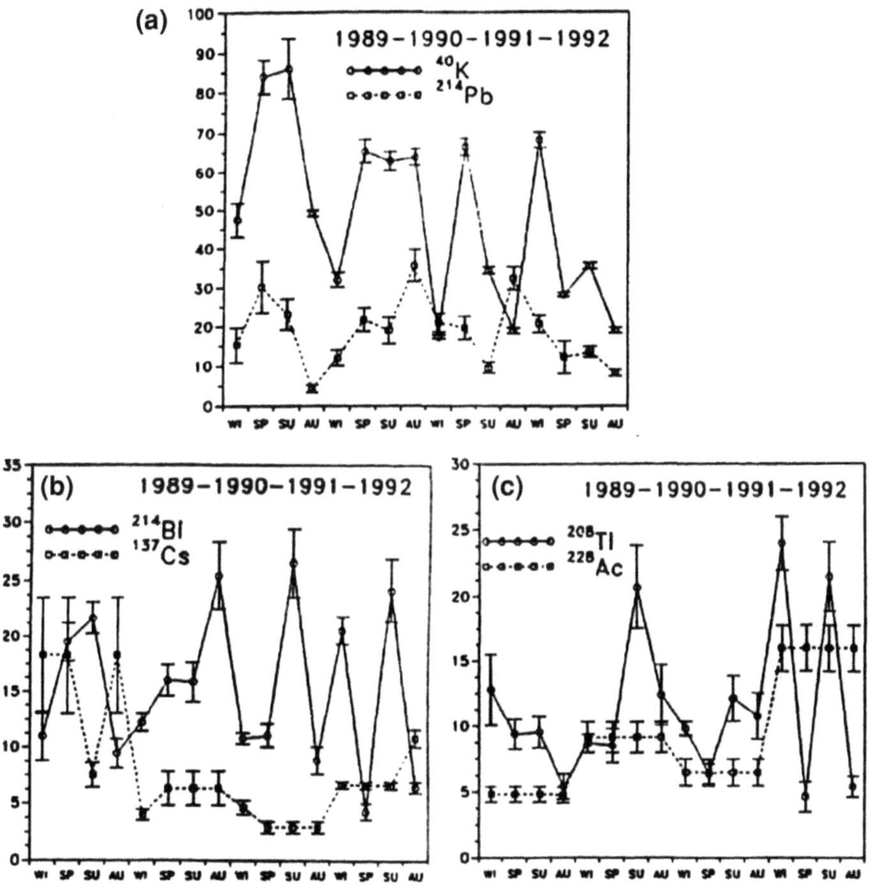

Fig. 3.15 a–c Seasonal variation of radioactivity during 1989, 1990, 1991, 1992 for some radionuclides

Table 3.9 Observed decay chain of the radionuclides and their min and max concentration

Nuclide	Decay chain[a]	Half-life	Concentration[b] (mBq/m^3)	
			Minimum	Maximum
^{214}Pb	^{238}U	26.8 m	13.66 ± 1.82	21.97 ± 4.83
^{208}Tl:	^{232}Th	3.1 m	8.65 ± 1.50	14.93 ± 2.16
^{214}Bi	^{238}U	19.9 m	12.36 ± 1.85	18.95 ± 2.21
^{137}Cs		30 y	4.34 ± 0.41	15.97 ± 2.02
^{228}Ac	^{232}Th	6.13 h	4.84 ± 0.82	12.74 ± 1.50
^{40}K		1,27.10^{9}y	34.25 ± 1.54	58.75 ± 4.06

[a] Decay chain in which the nuclide appears

[b] Mean values of the measurements for the four years

Table 3.10 Mean values of the air–temperature, relative humidity, atmospheric pressure, rainfall & wind speed in Thessaloniki for 1989, 1990, 1991, 1992

Months	Temperature (°C)	Humidity (%)	Wind speed[a] (ms^{-1})	Atm. press. (mmHg)	Rain (mm)
Jan	5	80	3.4	786.7	47
Feb	7	77	3.2	768.3	39
Mar	10	72	2.7	768.5	42
Apr	15	67	2.8	758.2	52
May	20	68	2.4	763.1	54
Jun	24	62	3.3	760	43
Jul	26	60	3.8	760.3	44
Aug	26	58	3.6	762.1	30
Sep	23	66	3.3	765	42
Oct	18	70	2.7	760	58
Nov	12	79	2.9	765.7	63
Dec	7	80	3.1	768	60

[a] The wind directions on the average for 1989, 1990 and 1992 are: Northern 15%, North-Eastern 6%, Eastern 5%, South-Eastern 4%, Southern 6%, South-Western 7%, Western 3%, North-Western 5% and Calm 49%

Conclusion

1. The most frequently occurring radionuclides in the atmosphere of Thessaloniki in 1989-1992 are: ^{214}Pb, ^{208}TI, ^{214}Bi, ^{137}Cs, ^{228}Ac and ^{40}K.
2. A correlation was found between the atmospheric radioactivity and the air parameters–speed, humidity and temperature.

3.1.4 Study on the Diffuse Radioactivity and Radionuclides Identification in the Major Area of Sindos—Thessaloniki Before and After the Chernobyl "Cloud" Appearance

The purpose of these studies was to compare the radioactivity levels before and after the Chernobyl cloud appearance as well as to identify the radionuclide contained in the cloud.

The daily values of the radioactivity level for 1981–1987 are measured by detectors at 3.5 m in height over the ground in counts per minute (cpm) [182].

The data obtained from measurements, carried out in May, 1986, show the radioactivity level in the air, in the grass and on concrete (Fig. 3.13). The lower dotted line illustrates the radioactivity degrees for previous years (1981–April 1986).

The background radioactivity for the period 1981–1986 has not exceeded the level of 25 cpm (Fig. 3.16), except for August, when there are measured slightly higher values because of the increased cosmic radiation, The value of 25 cpm can be assumed as a characteristic value for the Greek industrial area, where there are

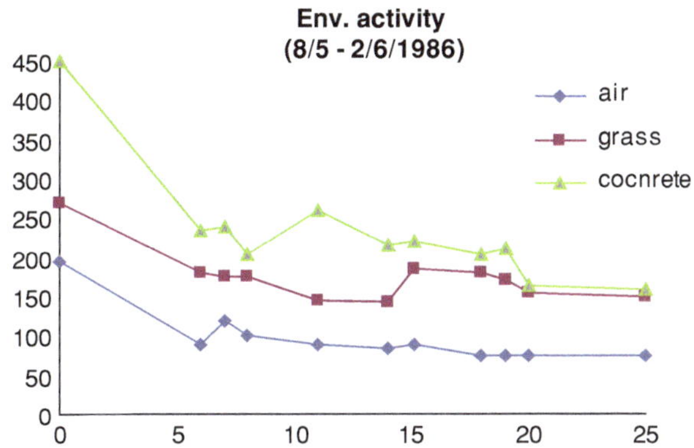

Fig. 3.16 Environmental radioactivity from 08/05/1986 to 02/06/1986

no nuclear stations. During the first month after the Chernobyl cloud appearance, the air radioactivity values varied from 90 to 200 cpm (Fig. 3.16).

The radionuclides identified over the industrial area on May 12, 1986 are: ^{103}Ru, ^{106}Ru, ^{134}Cs, ^{137}Cs, ^{131}I. Their concentration ware estimated by application of the empirical relationship:

$$\frac{A/t_1}{S \cdot t_2} \cdot F = C$$

where A = total number of events, indicated with the letter A in the computer files; t_1 = analyzer, counting time in seconds, S = the rate of suction of the air pump in $m^3\ h^{-1}$, t_2 = the total time of suction measured in hours, F = the isotope conversion factor, C = concentration in Bq m^{-3},

^{103}Ru, half-life 39.6 days, C = 3.03 Bq m^{-3}
^{106}Ru, half-life 367 days, C = 1.73 Bq m^{-3}
^{134}Cs, half-life 2.04 years, C = 0.93 Bq m^{-3}
^{137}Cs, half-life 30 years, C = 1.03 Bq m^{-3}
^{131}I, half-life 8.06 days, C = 1.61 Bq m^{-3}

The found correlation coefficient R = 0.68 for a linear variation between the radioactivity and the relative humidity shows that humidity influences stronger the radioactivity in comparison to the other meteorological parameters.

Conclusions

1. With the arrival of the Chernobyl cloud on May 2, 1986 over Sindos, Thessaloniki, the diffuse radioactivity of areas with low radioactivity, due to absence of nuclear stations, was increased 400–800%.
2. The presence of ^{137}Cs with long half-life, in concentration of the same order in magnitude as those of other shorter half-life radionuclides, should be focused on monitoring the quantities of ^{137}Cs in living organisms and in food.
3. It was demonstrated that the increase in the relative humidity causes increase in the radioactivity level in air.

3.1.5 Radionuclide Diffusion in the Environment

The results from the measurements of the total column and the concentration of SO_2, O_3, NO_x in the industrial area of Thessaloniki [132] makes possible the estimation of the total charge and transfer of the air pollutants, taking into consideration the influence of the meteorological parameters in the industrial area of Thessaloniki [181, 191].

The direct correlation between the total column and the concentration of SO_2, O_3 and NO_x on the ground surface can be determined if we consider a well-mixed atmospheric layer [3, 25, 88, 132].

A linear proportionality has been found between the total column of *Cl* and the concentration of C(o) on the surface, for SO_2.

Therefore:

$$\begin{aligned} Cl &= \int_0^{\infty} C(z)dz, \\ h &= \frac{Cl}{C(o)}, \\ t_d &= \frac{h}{V_d} \end{aligned} \tag{3.1}$$

where:

h = equivalent height (depth) of mixing for SO_2 (in infinite column of a unique surface area) of the relating concentration of SO_2 on the surface;
t_d = "the mean half-life" of SO_2 for the time of dry deposition of the fallout and it is as well a quotient of the equivalent height h to the velocity v_d of the dry out of SO_2.

Measurements have been made of the intensity of the solar ultraviolet radiation in wavelength regions, where the absorptions spectra of SO_2, O_3, NO_x have special construction.

The calculation of the total column from the measured intensities is based on the work of A. W. Brewer as well as on the research team of the Atmospheric Pollution of Canada.

A model was developed describing in details the transport of the atmospheric radioactivity, taking in consideration the Gaussian distribution of the airborne radioactive substances. At present there have been attempts on applying the principles aforesaid (with the eventual changes if necessary) in order to correlate the calculation of the radionuclide diffusion in the environment with the concentrations of C(o), C(z), *Cl* and the above mentioned parameters h and t_d of active (photochemical contamination) and non-active air contaminants and aerosols from dispersed, consequent or surface sources, according to the local meteorological parameters.

3.1.5.1 Theoretical Aspects of the Diffusion Phenomena of Radioactive Substances in the Atmosphere

The purpose of the calculations of the diffusion phenomena of radioactive substances in the atmosphere is to estimate the concentration in the air layer exactly near the ground surface (the ground concentration corresponding to C(o) for SO_2) at a certain point of the region of a nuclear station [191]. It should be emphasized that for the estimation of radioactivity, the knowledge of the short time concentration of radioactive substances is less important than the influence of the total radioactivity in longer half-lives, e.g. one year. According to the so-called "diffusion factors", the soil radioactivity K can be calculated from its relationship with the rate of radioactivity Q and long-lived diffusion factor X (Fig. 3.14).

$$K = Q.X\left[\mathrm{Ci/m^3}\right] \tag{3.2}$$

The factor X depends on:

- the distance of the release point of the radioactivity from the measuring station;
- the weather conditions;
- the wind velocity;
- the height of the release point of the radioactivity;
- the several ways of deposition;
- the ground profile.

and is given from the following relationship:

$$X(x, y) = \frac{1}{\pi U \sigma_y(x)\sigma_z(x)} \cdot \exp\left(\frac{-H^2}{2\sigma_z^2(x)}\right) \cdot \exp\left(\frac{-y^2}{2\sigma_y^2(x)}\right) \tag{3.3}$$

where: X(x, y) = short time diffusion factor of the position (x,y) [s/m^3]
$\sigma_y(x)$ = horizontal diffusion parameter at the y—direction, perpendicular to the main diffusion direction (in the wind direction), [m]
$\sigma_z(x)$ = vertical diffusion parameter at the z—direction [m]
U = wind velocity [m/s]
x = distance between the measuring station and the point of release at the wind direction [m]
y = distance between the measuring station and the point of release perpendicular to the wind direction, [m]
H = effective height of the release point.

The diffusion parameters σ_y and σ_z are defined by Pasquill [16] through the meteorological parameters and the ground profile of the region, which are the main factors that influence the level of radioactive diffusion in the atmosphere (Fig. 3.17).

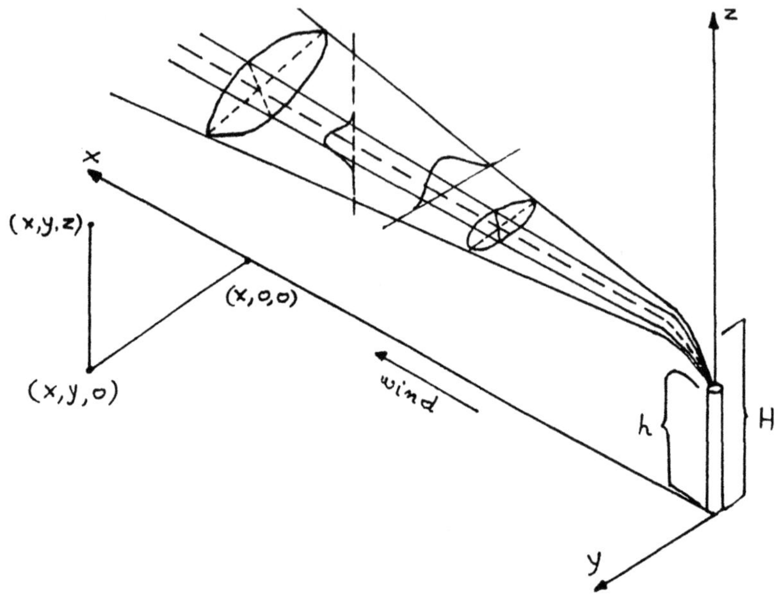

Fig. 3.17 Horizontal and vertical Gaussian distribution of airborne radioactive substances in the atmosphere [35]

Conclusions

The previous examinations made by Brewer Mark IV clearly show that the model aforesaid, illustrating the diffusion of radioactive substances in the atmosphere, must be modified in order to determine the ground concentration C(*o*) of SO_2.

This is a different consideration for the Gaussian distribution that is applied in the above model. Work on finding the correlation between the radioactive contaminants and non-radioactive (SO_2, O_3, NO_x) pollutants in the environment continues.

3.2 Transfer of Atmospheric Radioactivity in Soil and Plants

3.2.1 Natural Radioactivity of the Soils in Northern Greece

Natural and anthropogenic radioactivities in the soils are measured in order to find their sources and their way to man [107, 116, 119, 143, 155].

Here are presented updated data from experiments, the purpose of which is to explain the radionuclide contamination of the soil, plants and foods in the endangered regions of Greece and its surroundings.

Table 3.11 presents the specific activities and dose rates in Macedonia. Results are in agreement with other researchers findings [143]. The mean absorbed dose in air was found to be 20% higher than the mean dose rate of the whole country.

Table 3.11 Specific activities in the soil and absorbed dose rates in air for Macedonia

Specific activity (Bq kg^{-1})	Number of samples	Min dose	Max	Mean	Standard deviation
^{226}Ra	155	13.80	191.10	51.83	19.27
^{232}Th	156	9.60	135.50	51.66	20.59
^{40}K	154	132.70	1613.10	645.63	271.09
Dose equivalent (nGy h^{-1})	151	22.04	342.94	81.72	32.80

Table 3.12 Specific energy in soils and absorbed dose rates in air for Thrace

Specific activity (Bq kg^{-1})	Number of samples	Min	Max	Mean	Standard deviation
^{226}Ra	49	23.90	89.00	53.16	12.00
^{232}Th	50	24.80	84.10	53.62	13.35
^{40}K	50	292.50	1158.90	757.67	203.60
Dose equivalent (nGy h^{-1})	49	43.09	126.52	86.49	16.61

Table 3.13 Specific activities in soil and absorbed dose rates in air for Northern Greece

Specific activity (Bq kg^{-1})	Number of samples	Min dose	Max	Mean	Standard deviation
^{226}Ra	204	13.80	191.10	52.15	17.78
^{232}Th	206	9.60	135.50	52.14	19.07
^{40}K	204	132.70	1613.10	673.09	260.24
Dose equivalent (nGy h^{-1})	200	22.04	342.94	82.90	29.70

Table 3.14 Independent samples test for equality of ^{226}Ra means

	Region	Number of samples	Mean	Standard deviation	Standard error mean	Significance (2-tailed)
^{226}Ra	Macedonia	155	51.83	19.27	1.54	0.650
	Thrace	49	53.16	12.00	1.71	

Table 3.12 presents the specific activities and dose rates in Thrace. The results obtained for ^{40}K are in agreement with the previous measurements within the range of the standard deviation. The mean rate in air was found to be 25% higher in comparison with the mean value for the whole country.

Table 3.13 presents the specific activities in the soil and dose rates in Northern Greece. The mean absorbed dose rate in air was found to be 22% higher in comparison with the value for the whole country.

Tables 3.14 and 3.15 present independent samples t-tests for ^{226}Ra and ^{232}Th respectively. It was found in both cases that there are no statistically significant differences between Macedonia and Thrace for the Ra and Th means ($P > 0.05$).

Table 3.15 Independent samples test for equality of ^{232}Th means

	Region	Number of samples	Mean	Standard deviation	Standard error mean	Significance (2-tailed)
^{232}Th	Macedonia	156	51.66	20.59	1.64	0.438
	Thrace	50	53.62	13.35	1.88	

Table 3.16 Independent samples test for equality of ^{40}K means

	Region	Number of samples	Mean	Standard deviation	Standard error mean	Significance (2-tailed)
^{40}K	Macedonia	154	645.63	271.09	21.84	0.002
	Thrace	50	757.67	203.60	28.79	

Table 3.16 presents the same test for ^{40}K. It was found that there are statistically significant differences between Macedonia and Thrace for ^{40}K ($P < 0.05$). These results indicate the large variability of ^{40}K compared to that of the other two radionuclides.

Conclusions

The rural population in Northern Greece is exposed to at least 20% higher radioactivity doses compared to the population in the other regions of the country due to the presence of radionuclides in the soil. In Northern Greece, the presence of ^{137}Cs is twice higher compared to that in the other regions of the country. Taking in consideration the fact that in the next 50 years the external exposure absorbed dose from ^{137}Cs will be one order of magnitude lower than the average dose from the natural terrestrial radionuclides, one could accept that within the next 50 years, two rural residents of Northern Greece, roughly speaking, will receive the same dose as three rural residents from the other regions of the country due to the exposure to the combined action of ^{137}Cs and the natural radionuclides.

3.2.2 *Concentration Level of ^{137}Cs in Greece One Decade After the Chernobyl Accident*

The results of the measurements made are presented, concerning first: the level of ^{137}Cs ten years after the Chernobyl accident; second: comparative data between ^{137}Cs distribution in 1986 as well as its present distribution in the examined areas in Greece. The concentration of ^{40}K (0.0118% of the natural K) in the soils is examined as a ratio ^{137}Cs/^{40}K, although this ratio is not constant in the biological systems as the ratio Sr/Ca [52, 161, 182].

The studies of the distribution peculiarities of the long-lived radionuclides responsible for the gamma-radioactive contamination in the soils, that—regardless of the form of contamination, the type of soil and landscape—they are mainly concentrated in the top (0–5 cm). Table 3.17 shows the concentrations of ^{137}Cs

Table 3.17 Distribution of ^{137}Cs and ^{40}K as per prefectures in Greece

	Prefecture	Capital	Number of samples	Activity (kBq/m^2) ± S D	
				^{137}Cs	^{40}K
1	Achaia	Patra	8	1.8 ± 0.1	8.7 ± 0.2
2	Argolida	Nafplio	7	2.1 ± 0.1	9.9 ± 0.2
3	Arkadia	Tripoli	7	3.1 ± 0.2	8.8 ± 0.3
4	Arta	Arta	5	1.0 ± 0.1	9.4 ± 0.2
5	Attiki	Athens	10	4.0 ± 0.2	9.5 ± 0.3
6	Chalkidiki	Poligiros	11	3.0 ± 0.2	10.2 ± 0.2
7	Chios	Chios	4	0.4 ± 0.1	11.6 ± 0.3
8	Corfou	Corfou	8	3.1 ± 0.2	10.5 ± 0.1
9	Dodekanisa	Rodos	5	0.4 ± 0.1	12.9 ± 0.2
10	Drama	Drama	8	4.8 ± 0.3	11.6 ± 0.1
11	Etolokarnania	Mesologi	5	0.8 ± 0.1	12.9 ± 0.2
12	Evia	Chalkida	6	0.7 ± 0.1	11.9 ± 0.4
13	Evritanla	Karpenisi	7	1.7 ± 0.1	14.3 ± 0.2
14	Evros	Alexandroupoli	5	2.2 ± 0.1	15.8 ± 0.4
15	Florina	Florina	17	11.9 ± 0.3	8.2 ± 0.6
16	Fokida	Amfissa	8	1.6 ± 0.1	11.3 ± 0.7
17	Fthiotida	Lamia	3	7.6 ± 0.2	6.2 ± 0.4
18	Grevena	Grevena	7	10.6 ± 0.4	9.4 ± 0.3
19	Ilia	Pirogos	4	1.3 ± 0.1	9.9 ± 0.3
20	Imathia	Veria	15	12.7 ± 0.2	5.4 ± 0.1
21	Ioannina	Ioannina	14	3.1 ± 0.4	11.1 ± 0.4
22	Kefalonia	Argostoli	5	1.9 ± 0.1	10.7 ± 0.2
23	Karditsa	Karditsa	14	12.5 ± 0.4	6.7 ± 0.3
24	Kastoria	Kastoria	8	12.6 ± 0.5	6.6 ± 0.1
25	Kavala	Kavala	6	2.7 ± 0.1	11.8 ± 0.4
26	Kiklades	Ermoupolis	5	3.0 ± 0.2	12.8 ± 0.2
27	KILKIS	Kilkis	5	3.7 ± 0.1	10.0 ± 0.3
28	Korinthos	Korinthos	4	1.1 ± 0.1	10.1 ± 0.5
29	Kozani	Kozani	10	8.8 ± 0.3	11.7 ± 0.2
30	Kriti (Island)	Iraklio (l. town)	12	2.9 ± 0.2	10.7 ± 0.3
31	Lakonia	Sparti	5	2.0 ± 0.1	8.9 ± 0.4
32	Larisa	Larisa	12	9.4 ± 0.2	8.1 ± 0.2
33	Lesvos	Mitilini	3	1.0 ± 0.2	11.8 ± 0.3
34	Limnos (Island)	Mirina	5	3.5 ± 0.1	9.6 ± 0.5
35	Magnisia	Volos	12	7.9 ± 0.3	6.7 ± 0.4
36	Messinia	Kalamata	5	2.2 ± 0.2	10.9 ± 0.5
37	Pella	Edessa	5	10.0 ± 0.5	7.6 ± 0.2
38	Pieria	Katerini	18	14.4 ± 0.3	5.1 ± 0.1
39	Preveza	Preveza	5	1.4 ± 0.1	9.3 ± 0.2
40	Rodopi	Komotini	8	0.8 ± 0.1	10.1 ± 0.9
41	Serres	Serres	11	4.0 ± 0.1	14.5 ± 0.5
42	Sporades skopelos alonisos	Chora of Skiathos	8	3.0 ± 0.1	9.9 ± 0.3
43	Thesprotia	Igoumenitsa	6	2.8 ± 0.1	10.9 ± 0.3

(continued)

Table 3.17 (continued)

	Prefecture	Capital	Number of samples	Activity (kBq/m^2) ± S D	
				^{137}Cs	^{40}K
44	Thessaloniki	Thessaloniki	20	11.3 ± 0.4	7.8 ± 0.4
45	Trikala	Trikala	14	12.3 ± 0.6	7.9 ± 0.1
46	Viotia	Livadia	5	1.8 ± 0.1	10.2 ± 0.3
47	Xanthi	Xanthi	6	1.0 ± 0.1	16.5 ± 0.6

and ^{40}K. It becomes evident from the results that the fallout of ^{137}Cs from the Chernobyl "cloud" has a remarkable geographical variability and it is inversely proportional to the natural concentration of ^{40}K in soil. Cesium replaces potassium due to their equal dimensions and qualities [48].

Taking into consideration the composition of the Greek soils, most of which are rich in clay and silt and contain in excess exchangeable potassium, one would expect radiocesium contamination—up to about 5 cm in depth. On the other hand, the rate of transfer of radionuclides from ground to plants depends on the soil composition in the given area. Alkaline soils, which are rich in clay minerals (as Greek soils) retain cesium more stable than the acid ones, prohibiting its transfer to the plants and consequently to the animals and the humans.

It is not possible to achieve an accurate comparison between the present and the existing data of the distribution of ^{137}Cs in the soils of Greece because there are great differences in the distribution of radioactivity from place to place. The basic conclusion in the present work is that as a rule the same areas, which were considerably contaminated with ^{137}Cs in 1986, at present, they are found to be contaminated to the same order of magnitude but not below of 5 cm in depth.

The variety of typomorphic conditions on the territory of Greece, as well as the climatic conditions at the time of aerosol deposition, has been considerably influenced the ^{137}Cs behavior that reaches the soil. The behavior aforesaid depends on the type of the soil, its properties and cultivation. The ploughing of the cultivated soil has also been elaborated by the uneven distribution of ^{137}Cs and it hinders its determination in vertical and horizontal direction of spreading [129, 184].

Figure 3.18 shows ^{137}Cs concentration in comparison to that of ^{40}K. These results show that ^{137}Cs concentration is inversely proportional to that of ^{40}K in the soil (Figs. 3.19, 3.20, 3.21 and 3.22). Figure 3.23 presents comparative data of the logarithm of the concentration of ^{137}Cs and ^{40}K. The concentration is given by the following equation:

$$y = C[1 - A\exp(Bx)], \tag{3.4}$$

where:

$y = \log {}^{137}\mathrm{Cs}$
$x = \log {}^{40}\mathrm{K}$
A, B, C are constant coefficients, obtained by experimental data.

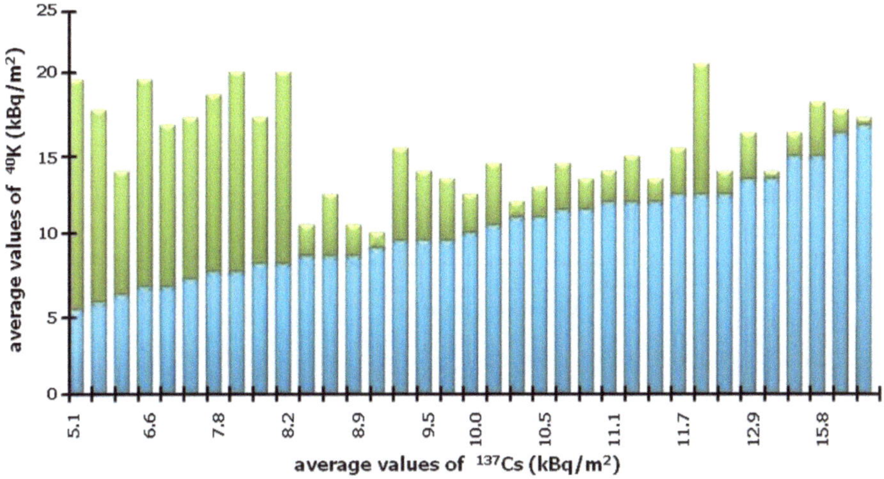

Fig. 3.18 The ^{40}K average concentrations versus that of ^{137}Cs (kBq/m^2)

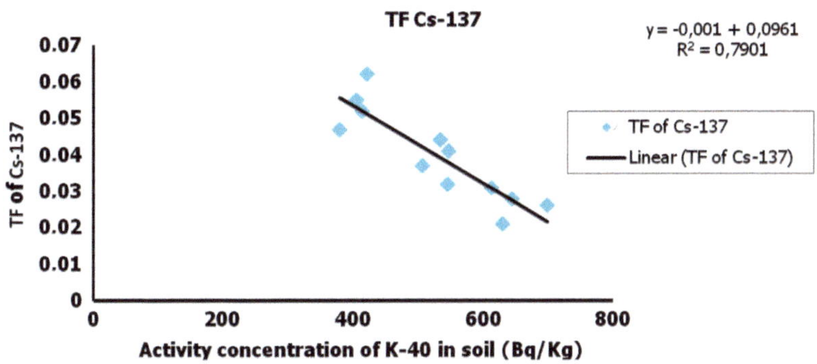

Fig. 3.19 Cs-137 transfer factor (TF) versus K-40 in soil

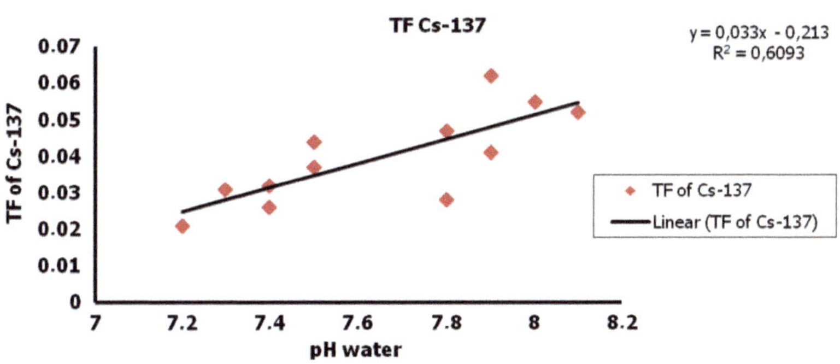

Fig. 3.20 Variation of transfer factor versus pH

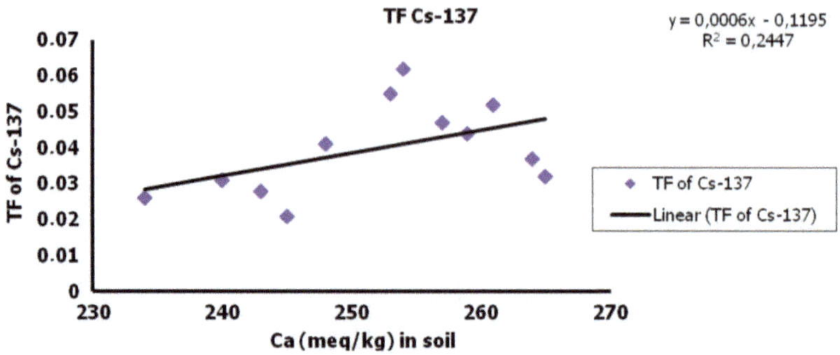

Fig. 3.21 Cs-137 transfer factor versus Ca++ in soil

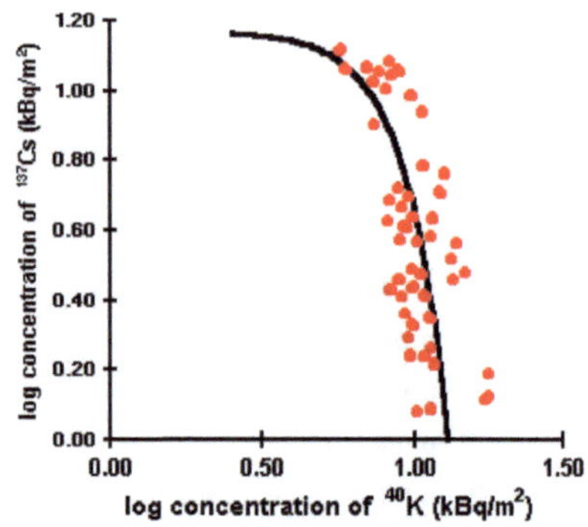

Fig. 3.22 The logarithm of ^{40}K concentration versus that of ^{137}Cs (kBq/m^2)

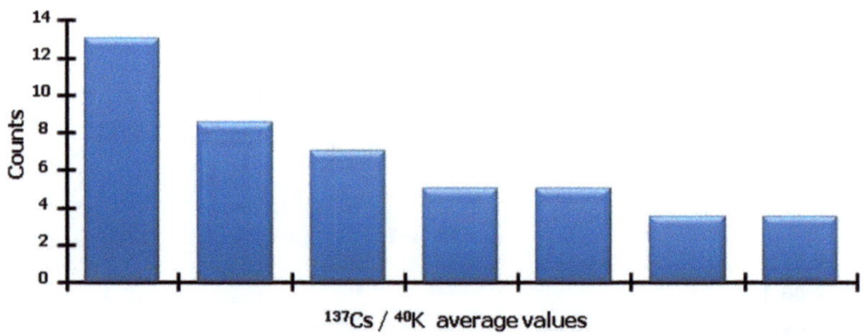

Fig. 3.23 The ratio ^{137}Cs/^{40}K from average values

The constant coefficients show that ^{137}Cs uptake from the soil is inversely proportional to the presence of ^{40}K in soil, in which there is a potassium deficiency Fig. 3.24 shows that the ratio ^{137}Cs/^{40}K is not constant, though cesium is congener to potassium.

The plants were separated from their roots and grains, thoroughly washed and cut to small pieces. They were air-dried for two days in the laboratory and then in

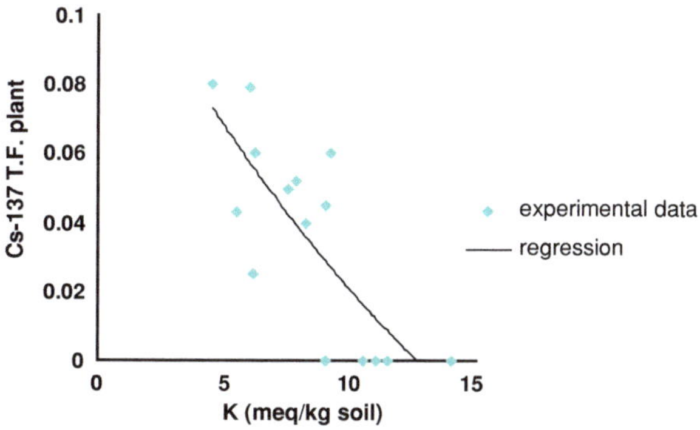

Fig. 3.24 ^{137}Cs transfer factor versus ^{40}K in soil

the oven (Melag) kept at 80°C. They were weighed (mass of dry plant) and burned at 500°C in another oven (Nabertherm) for at least 24 h to get carbon free white ash. This was done because Cs in a certain volume of ash is 8–9 times more than the Cs in the same volume of dry plant. It should be mentioned that Cs is not lost by oven-drying at 500°C (boiling points of Cs and K, 687 and >750°C). Plant ash was filled in the standard geometry white cup (cylindrical. 7 cm d. and 2 cm h.), recommended for measurement by the Greek Nuclear Research Centre–Demokritos.

Soil samples were also air-dried, oven-dried (80°C), pulverized, sifted (2 mm sieve) and weighed (mass dry soil). Then they were put to fill standard cups [161, 183, 224, 226].

The activity concentration of each isotope were calculated using the following equation:

$$A_s = C_s/\varepsilon P_r M_s \left(\mathrm{Bq\ kg^{-1}}\right)$$

where,

A_s is Activity concentration of the radionuclide (Bq kg^{-1})
C_s is counting rate of the gamma ray (counts per second)
ε is detector efficiency of the specific gamma energy
P_r is absolute transition probability of gamma decay
M_s is mass of the sample (kg)

Transfer factors (TF) are calculated as (As)plant/(As)soil meaning the activity concentration of the specific radionuclide in plant over their respective activity concentration in soil.

In average the Greek soil characteristic parameters are:

pH	7–8
Exchangeable K	0.5 meq/100 g
Clay	49%
Silt	37%
Organic matter	>5%

Table 3.18 Activity concentrations in Bq kg^{-1} dry plant and soil of Cs-137 and K-40

Samples	Activity Concentrations (Bq kg^{-1})		TF of Cs	Activity Concentrations (Bq kg^{-1})		TF of K
	Cs–137 plant	Cs-137 soil		K-40 plant	K-40 soil	
Grass 1	2.365	43	0.055	292.815	405	0.723
Grass 2	1.728	54	0.032	498.675	545	0.915
Grass 3	0.693	33	0.021	391.86	630	0.622
Corn 1	2.397	51	0.047	328.593	379	0.867
Corn 2	2.508	57	0.044	592.696	533	1.112
Corn 3	0.98	35	0.028	459.745	643	0.715
Wheat 1	2.914	47	0.062	342.694	421	0.814
Wheat 2	1.961	53	0.037	577.852	506	1.142
Wheat 3	0.936	36	0.026	372.732	698	0.534
Sugar-beet 1	2.288	44	0.052	330.372	414	0.798
Sugar-beet 2	2.009	49	0.041	555.282	546	1.017
Sugar-beet 3	1.24	40	0.031	421.668	612	0.689

TF = Transfer Factors from soil to plant

Table 3.19 pH variation of water and KCl from the dry plants

Plant samples	pH water	pH KCl
Grass 1	8	7.7
Grass 2	7.4	7.1
Grass 3	7.2	7.4
Corn 1	7.8	7.2
Corn 2	7.5	7.3
Corn 3	7.8	7.3
Wheat 1	7.9	7.5
Wheat 2	7.5	7.1
Wheat 3	7.4	7.4
Sugar-beet 1	8.1	7.4
Sugar-beet 2	7.9	7.2
Sugar-beet 3	7.3	7.2

Table 3.20 Concentration of exchangeable of exchangeable K, Ca and Mg in soil from the dry plants

Plant samples	K meq/kg soil	Ca meq/kg soil	Mg meq/kg soil
Grass 1	4.5	253	62
Grass 2	5.7	265	55
Grass 3	8.6	245	37
Corn 1	5.5	257	61
Corn 2	6.3	259	43
Corn 3	9.1	243	33
Wheat 1	6.3	254	64
Wheat 2	7.7	264	46
Wheat 3	9.5	234	39
Sugar-beet 1	6.2	261	59
Sugar-beet 2	7.1	248	41
Sugar-beet 3	9.4	240	32

Table 3.21 Concentration of clay, silt and sand in the respective soils collected from these areas

Soil samples	Clay (%)	Silt (%)	Sand (%)
1	45.7	41.2	26.2
2	37.6	33.4	21.6
3	27.9	30.3	37.7

Conclusions

1. ^{137}Cs concentration in Greek soils that has been registered in 1986 has not changed for a decade;
2. ^{137}Cs uptake from soil is inversely proportional to the presence of ^{40}K in the soil;
3. The ratio of ^{137}Cs and ^{40}K in the soil is not constant;
4. The soils are contaminated from 0 to 0.5 cm in depth. Alkaline soils, which are rich in clay and minerals (as the case with the soils in Greece is) retain cesium more stable than acid soils prohibiting its transfer in plants and therefore in animals and humans (Tables 3.18, 3.19, 3.20 and 3.21).

3.2.2.1 Radioactive Contamination in Greece due to the Chernobyl Accident and its Impact on the Environment

The results obtained are presented in Tables 3.22, 3.23 and 3.24.

^{137}Cs deposition in the soils in Greece varies between 0.4 and 14.4 kBq/m^2 (the values are up to May 1996) (Table 3.22). The standard deviations are low as the samples are taken from undisturbed soils at a small distance to one another [223, 225, 234, 240].

The study shows that Western Macedonia, Northern Thessalia and Central Macedonia are the most contaminated areas in the country. Depositions of the order of 3.5 and 12 kBq/m^2 are typical for 36% of the country [258, 264]. The above stated values are not corrected for ^{137}Cs during the nuclear tests in 1960s [244, 245]. These results are in compliance with the results of the total deposition of ^{134}Cs and ^{137}Cs (with activity less than 3% for 1996). The total cesium deposition measured in 1986 in Thessaloniki is 27.6 kBq/m^2 (with coefficient of variability-CV 55%). The fallout of ^{137}Cs in that city was approximately 40% of the ^{137}Cs fallout, which means that ^{137}Cs deposition was 16.5 kBq/m^2. Ten years later it has been reduced to 13.1 kBq/m^2 (CV 55%), which complies (with 10% deviation) of the measured value for 1996: 11.3 $\pm$ 0.4 kBq/m^2. The above mentioned deviation was due to differences in the methodology of samples, errors in measuring methodology, errors of the instruments, erosion, leaching, ploughing, mineralogical and chemical composition of soils, etc. The results comply with the

Table 3.22 Mean distribution of ^{137}Cs and ^{40}K per geographic region of Greece

Regions	Number of samples	^{137}Cs (kBq/m^2) ± SD	^{40}K(kBq/m^2) ± SD
Aegean islands	24	1.5 ± 0.2	11.3 ± 0.2
Dodecanisa	15	0.4 ± 0.1	12.9 ± 0.2
Eptanisa	26	2.6 ± 0.2	10.6 ± 0.1
Ipiros	45	2.2 ± 0.2	9.8 ± 0.2
Kiklades	19	3.3 ± 0.1	13.1 ± 0.3
Kriti	28	3.1 ± 0.2	9.9 ± 0.3
Macedonia	263	8.6 ± 0.3	8.9 ± 0.4
Peloponisos	87	2.8 ± 0.2	10.1 ± 0.3
Sporades	8	3.0 ± 0.1	9.9 ± 0.3
Sterea Ellada	106	1.8 ± 0.1	10.1 ± 0.3
Thessalia	80	9.2 ± 0.4	7.4 ± 0.3
Thrace	41	2.8 ± 0.2	10.8 ± 0.5

SD = Standard Deviation

Table 3.23 Concentration of ^{137}Cs and ^{40}K in corn and clover

Plant/soil type	C1 = Bq ^{137}Cs/kg dry plant	C2 = Bq ^{137}Cs/kg dry soil	CR1 = C1/C2	C3 = Bq ^{40}K/kg dry plant	C4 = Bq ^{40}K/kg dry plant	CR2 = C3/C4
Corn 1	2.4691	50.8487	0.0486	246.7070	421.0781	0.5859
Corn 2	3.1390	31.8405	0.0986	186.3712	305.2657	0.6105
Corn 3	3.0988	28.3729	0.1092	259.4918	388.3208	0.6682
Clover	5.5426	49.7160	0.1115	285.1879	288.3785	0.9889

Table 3.24 Transfer coefficient from ingested material to milk, meat and eggs of different animal species [285]

Ingested Material		Cs	Sr	I
Milk	Cow	7.9×10^{-3}	2.8×10^{-3}	1.0×10^{-2}
	Sheep	5.8×10^{-2}	5.6×10^{-2}	4.9×10^{-1}
	Goat	1.0×10^{-1}	2.8×10^{-2}	4.3×10^{-1}
Meat	Beef	5×10^{-2}	8×10^{-3}	4×10^{-2}
	Veal	2×10^{-1}	1×10^{-1}	–
	Mutton	1.7×10^{-1}	4×10^{-2}	3.0×10^{-2}
	Lamb	4.9×10^{-1}	3.3×10^{-2}	3.0×10^{-2}
	Goat	2.3×10^{-1}	2.6×10^{-3}	–
	Pork	2.4×10^{-1}	4.0×10^{-2}	3.3×10^{-3}
	Poultry	1.1×10^{-1}	8×10^{-2}	–
Egg	Poultry	4×10^{-1}	2×10^{-1}	3

data of rainfalls for the period 3—7 May, 1986 reporting rainfall height 10–20 mm in Northern Thessalia and in central Macedonia and 20–40 mm in Northern Macedonia, and less than 10 mm for the rest of the country [249, 265].

^{40}K concentration in soils in 1996 varies from 5.1 to 16.5 kBq/m^2 (Table 3.22). Higher values were observed in the islands, which was probably due to the geological characteristics of their soils, as ^{40}K is a natural radionuclide [251, 252]. The above mentioned values are in compliance with these, measured by other investigators. ^{40}K gives information about the total concentration of K in the soils. It is an element, which has the same size and properties as Cs. Areas with higher concentration of ^{40}K are less contaminated with ^{137}Cs, as there is enough potassium in the soils and that's why plans absorb less cesium [226, 230].

Table 3.23 presents the concentration of ^{137}Cs and ^{40}K in plants. Firstly corn and clover were studied. ^{137}Cs concentration in corn was 2.47 Bqkg^{-1} dry plants, and in clover was 5.54 Bq kg^{-1} dry plants. The concentrations per kilogram of fresh plant are less than 1 Bq. The transfer factor of ^{137}Cs CR = Bq ^{137}Cs kg^{-1} dry plant, Bq ^{137}Cs kg^{-1} dry soil is 0.0486 for corn and 0.1115 for clover. For ^{40}K the transfer factors are 0.5859 for corn and 0.9889 for clover. The results aforesaid show that the plants in spite of the high uptake of ^{40}K, they uptake some Cs as well [242, 243]. The results for the corn comply with these from 1987, which are: CR = 0.037 ± 0.015 (upper limit 0.052). The CR of clover is similar to that of grass (0.129). It is evident from the results from 1987 that the concentration of the total K in the soil influences the uptake of Cs [138, 139].

3.2.2.2 Transfer of ^{137}Cs and ^{40}K from Soil to Plants in Northern Greece

The purpose of these studies was to make clear the transfer and in details of ^{137}Cs and ^{40}K from soil to plants in Northern Greece [250, 256].

The results obtained are presented in Tables 3.25, 3.26 and 3.27.

Corn was studied. ^{137}Cs activities vary between 0.01 and 2.13 Bq/kg and in soils between 23.1 and 51.1 Bq/kg. The transfer factors vary from 0.0015 to 0.0771. The transfer factor values comply with these, which were found by another research group ten years earlier [11]. It means that the fixing mechanism of Cs in the soil has not affected Cs presence and its uptake from plants.

In case of increasing ^{40}K concentration in plants, Cs concentration decreases. The correlation coefficient is 0.66 for a linear fit versus to ^{40}K in the soil. The studies show that plants, which have been grown in soils, which are poor in K (airport case), are expected to uptake more Cs [261].

Table 3.26 shows the variation of pH of water and KCl.

The pH varies from 7.53 to 8.32 for water and from 7.08 to 7.56 for KCl respectively. The low values of R show that pH under 8.4 does not influence the transfer factors (TF). This is in compliance with the results from other studies as well [222].

We could conclude that ^{137}Cs uptake from corn in 1998 has not changed in comparison with the situation in 1988. It seems that the fixing of ^{137}Cs in the soil does not influence the uptake to the extent it had been initially anticipated. The present work must be repeated for the years 1999 and 2000, in order to establish a long-term model of forecasting contamination or to verify the UNSCLEAR model and evaluate its coefficients.

Table 3.25 Activities (Bq kg^{-1}) of ^{137}Cs and ^{40}K in dry plant and soil and transfer factors from soil to plants

Place	Plant		Soil		Transfer factor	
	^{137}Cs	^{40}K	^{137}Cs	^{40}K	^{137}Cs	^{40}K
Agios 1	0.07	428	42.4	549	0.0015	0.780
Agios 2	0.02	391				
Agios 3	1.59	499	40.0	582	0.0399	0.856
Agios 4	0.00	333	51.1	531	0.0000	0.628
Agios 5	0.01	314	41.5	530	0.0003	0.593
Airport 1	1.68	297	38.3	199	0.044	1.488
Airport 2	2.10	251	27.8	219	0.075	1.146
Airport 3	0.90	337	39.8	211	0.023	1.593
Airport 4	2.06	246	34.5	204	0.060	1.203
Airport 5	2.13	233	27.6	234	0.077	0.998
Drama 1	1.17	352	23.6	545	0.050	0.646
Drama 2	1.13	321	23.6	464	0.048	0.692
Drama 3	1.16	411	25.9	430	0.045	0.956
Drama 4	0.97	405	26.2	466	0.037	0.868
Drama 5	1.34	423	23.1	472	0.058	0.896

Table 3.26 pH variation

Place	pH water	pH KCl
Agios 1	7.6	7.3
Agios 2	7.5	7.2
Agios 3	7.8	7.3
Agios 4	7.6	7.1
Agios 5	7.8	7.4
Drama 1	7.7	7.2
Drama 2	7.6	7.2
Drama 3	7.8	7.2
Drama 4	7.8	7.3
Drama 5	7.6	7.1
Airport 1	8.3	7.6
Airport 2	8.2	7.5
Airport 3	8.2	7.5
Airport 4	8.1	7.5
Airport 5	8.3	7.5

Activities of Cs-137 and K-40 kg^{-1} dry are given in Table 3.25. Cs-137 activities in the plant vary between 0.01 and 2.13 Bq kg^{-1} and for the soil between 32.1 and 51.1 Bq kg^{-1}. Transfer factors vary between 0.0015 and 0.0771 and are in agreement with the values found 10 years ago by others researches [224], which means that the fixation mechanism, in the soil was not really affected by the availability of Cs and its uptake by the plant.

Table 3.27 Concentration of exchangeable K, Ca and Mg in the soil

Place	K meq/kg soil	Ca meq/kg soil	Mg meq/kg soil
Agios 1	11.0	240	29
Agios 2	10.0	225	39
Agios 3	9.3	242	33
Agios 4	14.0	240	32
Agios 5	11.0	244	27
Drama 1	7.8	248	15
Drama 2	7.3	244	16
Drama 3	9.2	260	16
Drama 4	8.5	255	17
Drama 5	9.4	257	16
Airport 1	5.4	261	62
Airport 2	5.9	267	72
Airport 3	6.2	263	64
Airport 4	6.1	256	63
Airport 5	4.6	255	60

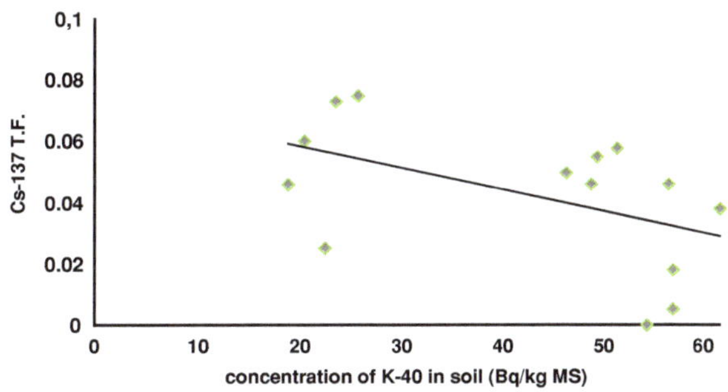

Fig. 3.25 Cs-137 transfer factor versus K-40 in soil regression; • experimental data

The Cs-137 concentration in the plant decreases as the concentration of K-40 in the plant increases. The correlation coefficient is −0.66 for a linear fit. Figure 3.25 shows a similar correlation of Cs-137 transfer versus K-40 concentration in the soil. This finding means that plants grown in the soils poor in K are expected to have higher Cs uptake.

Table 3.28 shows the sequence of calculations used to find the mean *p* coefficients of the UNSCEAR model for the year 1998 which vary between 0.034 and 00.112 and are in agreement with the previous works [223]. This finding supports the fact that Cs-137 availability from soil to plant in 1998 is similar to the 10 years ago (1988).

Figure 3.26 presents a lignite spectrograph showing the Th-234 and Ra-226 peaks. The information of the peak report was used to find the concentrations of U-238 (60 Bq kg^{-1}) and Ra-226 (95 Bq kg^{-1}) for the particular sample. These concentrations are not insignificant and further work is needed to find

Table 3.28 Mean p coefficients of the UNSCEAR model

a/a	Samples ID	Location	F_o (1996) mCi/km^2	Mass (g) dry plant	C_m (CU) (1998)	P_{1998} C_m/F_o	Mean P $_{1998}$	
1	Agios Ath	a	Thessaloniki	305	166.55	8.62110	0.02836	0.034
2	Agios Ath	b	Thessaloniki	305	145.90	9.33143	0.03059	0.034
3	Agios Ath	c	Thessaloniki	305	144.19	14.29599	0.04687	0.034
4	Agios Ath	d	Thessaloniki	305	153.81	0.00000	0.00000	0.034
5	Agios Ath	e	Thessaloniki	305	165.14	9.67446	0.03172	0.034
6	Airport	f	Thessaloniki	305	137.05	18.08903	0.05931	0.062
7	Airport	g	Thessaloniki	305	112.17	20.39474	0.06687	0.062
8	Airport	h	Thessaloniki	305	125.01	14.06224	0.04611	0.062
9	Airport	i	Thessaloniki	305	114.34	20.39474	0.06687	0.062
10	Airport	j	Thessaloniki	305	135.82	21.65127	0.07009	0.062
11	Drama	1	Drama	130	92.136	15.06904	0.11592	0.112
12	Drama	2	Drama	130	97.036	15.34859	0.11807	0.112
13	Drama	3	Drama	130	114.24	14.13992	0.10877	0.112
14	Drama	4	Drama	130	105.88	13.56686	0.10436	0.112
15	Drama	5	Drama	130	119.47	14.57005	0.11208	0.112

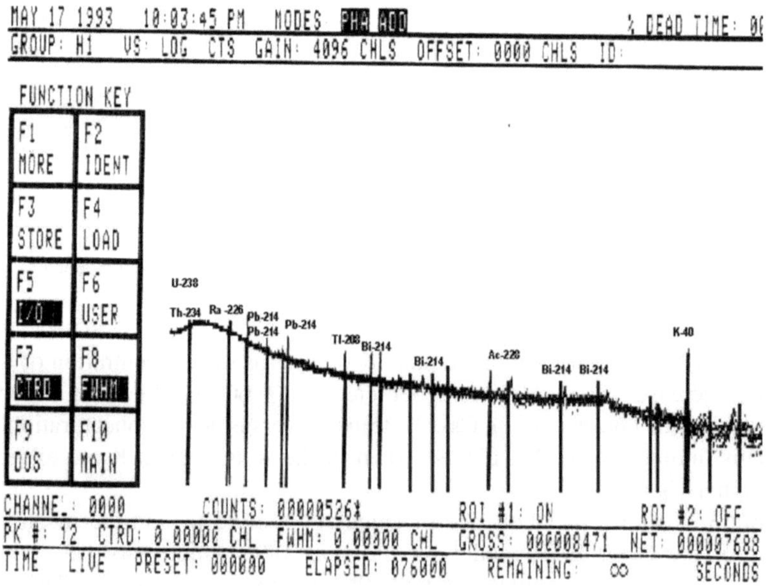

Fig. 3.26 Lignite spectrograph

concentration in large numbers of lignite samples and possibly use air samplers to trap fly ash from thermoelectric plants and proceed with its analysis for natural radionuclides.

3.3 Transfer of Radioisotopes in Food Stuffs

3.3.1 Mechanism of the Transfer of ^{121}I and ^{137}Cs from Cow Milk to Cheese and other By-Products

3.3.1.1 Study of the Iodine Transfer from Milk to Cheese and to Other Milk Products

The studies were carried out on batches of cheese produced under equal conditions and each cheese batch had a different level of ^{131}I contamination [149, 150, 175, 186–188]. The milk, which is used for the production of feta cheese and mytzithra, has the following mean composition:

Fat	3.9 (%)
Proteins	3.7
Lactose	4.7
Ash	0.7
Total solids	13.0

The pH value of the milk is 6.64. The produced curd and cheese whey have the following approximate composition:

	Cheese curd (%)	Cheese whey (%)
Fat	20.3	0.9
Protein	16.9 (N.P.N. 1.5)	1.1
Lactose	1.5	5.2
Ash	2.0	0.5
NaCl	2.5	–
Total solids	43.2	7.7

The yield of cheese from milk was approximately 16.0%

The mytzithra and its by-product lactose serum have the following approximate composition:

	Mytzithra (%)	Lactose serum (%)
Fat	15.1	0.0
Protein	12.9	0.2
Lactose	7.0	5.1
Ash	2.0	0.7
Total solids	37.0	6.0

The yield of mytzithra from milk is 8.5%

Studies were carried out on two batches of feta cheese with different contamination level [103, 104, 186–189, 223]. The milk used has the following approximate composition:

Fat	4.3 (%)
Proteins	4.1
Lactose	5.2
Ash	0.8
Total solids	14.4

The pH of milk is 6.65

The cheese curd and the cheese whey produced have the following approximate composition:

	Cheese curd (%)	Cheese way (%)
Fat	21.5	0.7
Protein	19.0 (N.P.N. 15%)	1.0
Lactose	1.9	5.9
Ash	2.2	0.5
NaCl	2.7	–
Total solids	47.3	8.1

The yield of cheese from milk was 18.4%. The approximate composition of mytzithra and its by-product lactose serum was:

	Mytzithra (%)	Lactose Serum (%)
Fat	14.5	0.0
Proteins	12.5	0.3
Lactose	7.0	5.2
Ash	2.0	0.7
Total solids	36.0	6.2

The yield of mytzithra from milk is 10.0%

The results show that when processing the contaminated milk into cheese and curd, the nuclides ^{131}I and ^{137}Cs, used in the study, pass into the aqueous whey, lactose serum). The measurements made by the γ-spectroscopy showed that the concentration of ^{137}Cs in the water phase was higher than that of ^{131}I, i.e. 82.4% for ^{137}Cs and 71.1% for ^{131}I. This was probably due to iodine tendency to participate in chemical covalent bonds with organic substances which inhibits its transfer into aqueous solutions in ionic form and from the cesium's position in first group of the periodic table [13, 194]. In case when the contamination level in milk increases, bigger amount of ^{131}I and ^{137}Cs is accumulated in cheese curd. However, the amount of the transferred radioactivity to cheese whey is 75.0% in case of lowest contamination value and 68.5% in case of highest value of

Table 3.29 Distribution of ^{131}I from cow milk to cheese curd and cheese whey

Samples	Radioactivity (Bq/kg)[a]					Radioactivity transferred from milk to cheese curd whey[b] (%)					
Production lines	I	II	III	IV	V	I	II	III	IV	V	Average values
Milk	542	1079	1481	2100	4102						
Curd	853	1893	2818	3524	8030	25	28	30	30	31.5	28.9
Whey	480	922	1216	1713	3346	75	72	70	70	68.5	71.1

[a] Average values from the five experiments made
[b] Calculated considering the yield

Table 3.30 Presence of ^{131}I in mytzithra and lactose serum

Samples	Radioactivity (Bq/kg)	Radioactivity, transferred from the starting material to the products[a] (%)
Starting material	5873	100
Cheese whey (90%) and milk (10%)		
Products		
a) Mytzithra	11805	17.1
b) Lactose serum	5322	82.9

[a] Calculated concerning the yield

Table 3.31 Distribution of ^{137}Cs from cow milk to cheese curd and cheese whey

Samples	Radioactivity (Bq/kg)[a]			Radioactivity transferred from milk to cheese curd and cheese whey[b] (%)			
Production lines	I	II	III	I	II	III	Average values
Milk	1960	3905	5828	100	100	100	
Milk products							
a) Cheese curd	1696	3750	6120	15.9	17.7	19.3	17.6
b) Cheese whey	2020	3940	5762	84.1	82.3	80.7	82.4

[a] Average values from the five experiments made
[b] Calculated considering the yield

^{131}I contamination. These values are 84.1 and 80.7% for ^{137}Cs respectively (Tables 3.29, 3.30, 3.31, 3.32).

The lactose serum that is obtained in case of mytzithra production has higher presence of radioactive ^{131}I compared to that of the cheese whey with values 82.9 and 71.1% respectively. In case of ^{137}Cs contamination, we have the opposite result and values 78.9 and 82.4% respectively. Taking into consideration these results, it is evident that the distribution of both used radioactive isotopes depends on the isotope's nature, on the composition of the starting material (milk) and on the production of milk products. The dissimilar distribution of the same isotope in cheese curd and mytzithra can be explained by the different composition

Table 3.32 ^{137}Cs distribution in mytzithra and lactose serum

Samples	Radioactivity (Bq/kg)	Radioactivity, transferred from the starting material to the products[a] (%)
Starting material		
Whey (90%) and milk (10%)	5830	100
Products		
a) Mytzithra	12330	21.1
b) Lactose serum	5108	78.9

[a] Calculated considering the yield

characteristics of the raw material used as well as by the differences in technological methods applied in mytzithra production. In this case the heating processing, which is applied, is of great importance as it influences the transfer of radioactive isotopes to the water phase [188].

Conclusions

The results show that the radioisotopes ^{131}I and ^{137}Cs follow the aquatic phase, during the processing of the contaminated milk. It has been accepted that in the course of time the concentrations of radioactivity in cheese have been changing because of using radionuclides with different half-life. The results presented in this study are useful for assessing health risks and for working out the public health policy for endangered areas of receiving radioactive fallout in the future. In addition, this information enables the producers to select parameters of processes, which will reduce to a minimum the transfer of radioactive contaminants in products.

3.3.2 Mechanism of ^{131}I Transfer from Feta Cheese After Immersion in Water and in Brine

The experimental data concerning ^{131}I release from Feta cheese are summarized in Tables 3.33, 3.34 and 3.35, in zone A, B, C and D respectively. All the measurements refer to the ratio C: Co, where C is the final and Co the initial concentration of ^{131}I, measured in cps before and after immersion in different solutions. Table 3.35 presents the results from ^{131}I penetration in feta cheese after immersion in different solutions, contaminated with ^{131}I.

It is evident that ^{131}I has transferred from feta cheese into brine, the amount of the transferred radionuclides depending mainly on the time of immersion in brine. Most of the radionuclides are transferred from the surface to the inner parts of cheese curd [186–189].

Neutral salts such as NaCl and KCl are used in the present study in order to determine whether the level of penetration and release of radionuclides is influenced by the presence of Na^+, K^+ or Cl ions, which are used in the production of

Table 3.33 Ratio of the concentration of ^{131}I in the cheese zones before and after immersion in a solution of NaCl and water

Zone	Immersion solution	5 h[a]		24 h[a]		48 h[a]	
		C:Co[b]	C:Co[c]	C:Co[b]	C:Co[c]	C:Co[b]	C:Co[c]
A	NaCl	0.743	0.706	0.788	0.708	0.727	0.686
B	NaCl	0.796	0.728	0.484	0.536	0.337	0.345
C	NaCl	0.869	0.892	0.666	0.609	0.461	0.377
D	NaCl	0.974	0.967	0.710	0.673	0.585	0.457
A	H_2O			0.584		0.602	
B	H_2O			0.535		0.413	
C	H_2O			0.649		0.448	
D	H_2O			0.728		0.643	

[a] Immersion time
[b] Pressure 5.5 kNm^{-2}
[c] Pressure 12.3 kNm^{-2}

Table 3.34 Ratio of the concentration of ^{131}I in the cheese zones before and after immersion in NaCl and KCl solutions and water

Zone	16 h[a]			24 h[a]			48 h[a]		
	Solution			Solution			Solution		
	NaCl	KCl	H_2O	NaCl	KCl	H_2O	NaCl	KCl	H_2O
A	0.447	0.423	0.511	0.380	0.418	0.400	0.469	0.415	0.407
B	0.520	0.414	0.532	0.313	0.346	0.379	0.290	0.269	0.361
C	0.607	0.556	0.629	0.444	0.452	0.477	0.346	0.385	0.423
D	0.590	0.535	0.758	0.576	0.678	0.706	–	–	0.459

[a] Immersion time

Table 3.35 Correlation of ^{131}I concentration before and after immersion in solutions of NaCl and KCl

Zone	24 h[a]		48 h[a]	
	Solution		Solution	
	NaCl	KCl	NaCl	KCl
A	0.516	0.494	0.370	0.492
B	0.601	0.633	0.344	0.516
C	0.664	0.633	0.469	0.594

[a] Immersion time

cheese. The nature of the used solution does not influence greatly the amount of removal or penetrated ^{131}I. Usually there is a variety of products containing potassium with sodium on the market. But replacement of potassium by sodium must be done carefully, as potassium is also a physiological electrolyte, which has a function of stimulating cardiac and renal function [193].

A greater amount of ^{131}I penetrates in cheese curd in comparison with the amount, which is removed in the serum. It seems that the transport mechanism of ^{131}I is an

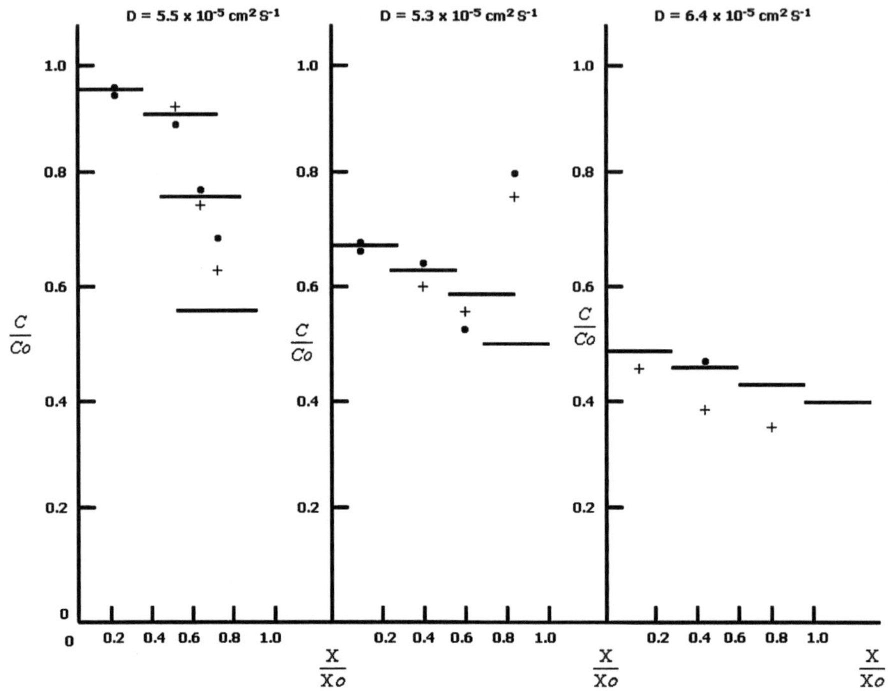

Fig. 3.27 Experimental data from Table 3.33 for different pressures

elaborate and dynamic process that delays its exit and facilitating its entry in the cheese curd. In order to explain the mechanism of ^{131}I transfer, we accept that:

a. The ^{131}I movement in feta cheese is due to a gradient of ^{131}I concentration in the cheese-salt solution system;
b. The ^{131}I moves from cheese to brine due to the fact that the brine volume is much greater than this of the cheese.

These two statements together with the constant diffusion coefficient (Batalas, 1986) of ^{131}I have been used for calculating the theoretical average values of ^{131}I concentration in each zone. The experiments and the theoretical values of ^{131}I concentration are very close as Fig. 3.27 shows.

Figure 3.27 presents experimental data from Table 3.33 of different pressure; dot (•) represents 5.5 kNm^{-2}, cross (+) represents 12.3 kNm^{-2} and the straight line represents the theoretical constant D. The letters a, b, c stand for 5, 24, 48 h immersion time respectively. The ratio X:Xo is a dimensionless parameter where Xo is the thickness of the undivided piece of cheese from the center and X is the distance of the middle point of each zone (A, B, C, D) from the center of the piece of cheese.

The diffusion constant varies from 5.5×10^{-5} to 6.4×10^{-5} cm^2s^{-1}. There is a very satisfactory agreement between the experimental and the theoretical values of

Table 3.36 Ratio of ^{131}I concentrations in the cheese samples before and after immersion in contaminated (10 kBq kg^{-1}) solutions of NaCl, KCl and H_2O, equal to the cheese sample mass

Zone	NaCl[a]			KCl[a]			H_2O[a]
	Concentration (%)			Concentration (%)			
	10	18	25	10	18	25	
A	0.841	0.848	0.861	0.839	0.833	0.852	0.904
B	0.737	0.744	0.723	0.740	0.775	0.726	0.835
C	0.622	0.662	0.601	0.646	0.708	0.628	0.773

[a] 48 h immersion time

^{131}I concentration, which is more obvious for a shorter time of immersion rather than for longer time of immersion. In the last zone (the outer part) of the cheese curd a deviation is observed from the above-mentioned statement. The fact that the deviation from the theoretical values of concentration depends on the immersion time has been of interest. The aforesaid specificity of the outer layer of the cheese is due to the fact that feta cheese remains unchanged during the experiment in comparison with the rest of its contaminated sectors in the brine. In the course of time a part of the released ^{131}I begins penetrating in the cheese where the concentration of ^{131}I due to the flow from feta cheese to the brine is not the same for the outer portion of the cheese, except for the first five hours.

In the present study the emphasis is on the mechanism rather than on the description of ^{131}I transfer from feta cheese to the brine. Therefore, the differences in concentration between theoretical and experimental values, especially for the last cheese zone, were not practically useful for our purposes.

Using the same model in explaining ^{131}I penetration in feta cheese from the brine, it was found (Table 3.36) that the diffusion coefficient of ^{131}I penetration was 2–3 times higher compared to the coefficient of removal of ^{131}I from feta cheese into brine. It seems that the removal and penetration of ^{131}I are two distinct processes and their course depends on different factors. All the aforesaid observations lead to the point that the protein structure of cheese plays an important role in the mechanism of transport of ^{131}I and therefore it shall be taken in consideration upon description of the whole process. Figure 3.28 attempts at a schematically presentation of the native inner structure of feta cheese.

The protein matrix of the cheese is relatively loose and porous in structure. According to our proposed model, ^{131}I moves freely along the cheese macropores interacting continuously with the bulk material. During this interaction part of ^{131}I is moving through the bulk material where it is practically trapped in the protein matrix but at the same time another part is released from the bulk material to macropores and is transferred outward without great difficulty. According to this model, the amount of free or trapped ^{131}I depends directly on the state of proteins due to the presence of NaCl or KCl.

In the first hours of immersion in the brine, the ^{131}I, which is supposed to be located in cheese macropores, moves freely outwards without any interaction with the cheese bulk material. Only this step of the whole process can be described by a constant diffusion coefficient. The amount of ^{131}I that leaves the cheese bulk

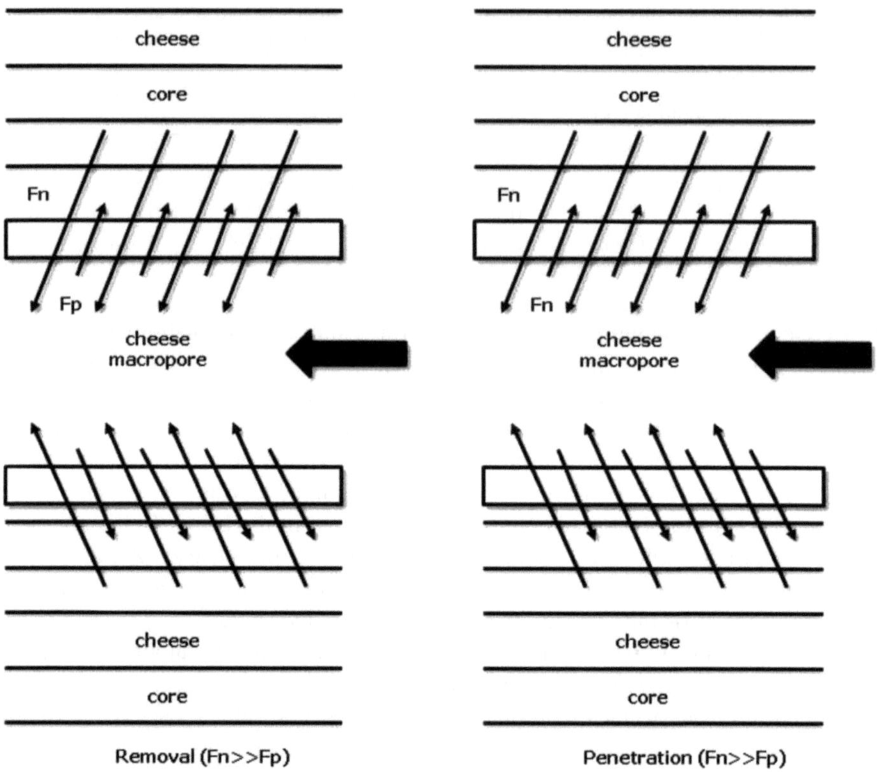

Fig. 3.28 Schematical presentation of the inner structure of cheese and the process of ^{131}I transfer. (F_n is the flux of ^{131}I from the matrix macropore and F_p is the flux of ^{131}I from the macropore to the matrix)

material later to the macropores (Tables 3.33, 3.34 and 3.35) is due to the ^{131}I interacted with the protein matrix during its movement outwards.

The higher the salting the faster the ^{131}I removes. In case of leaving the cheese into brine, the NaCl concentration is reduced. It means that additional amount of salts is transferred from brine to cheese. The elevation of salt concentration plays a considerable role in changing the native state of cheese proteins and therefore controlling the amount of ^{131}I removal.

Since solubility of proteins in water depends on clustering of water molecules around the hydrophilic ionic groups, removal of water molecules to other ions will decrease protein solubility and, therefore, create better conditions for ^{131}I removal from cheese bulk material and macropores.

If the transport mechanism was a simple diffusion, with the diffusion coefficient D being constant, as indicated in Fig. 3.27, in 48 h equilibrium should be established in the system: salt solution–cheese, which means that the concentration of ^{131}I trapped inside the cheese matrix should be constant. It is obvious that this is not the case as Fig. 3.27 shows, where the concentration of ^{131}I in the middle zone is lower compared to the calculated concentration and much higher in the outer zone (considering D is constant).

These findings emphasize the fact that the removal of ^{131}I depends mainly on structural changes of the protein matrix because of the presence of NaCl or KCl. The same results are also found when the brine is replaced by water with the rate of ^{131}I removal found to be slower, due to the fact that the native protein structure was not affected.

In the case of ^{131}I penetration from the contaminated brine to the cheese curd, the ^{131}I follows the aqueous phase entering the cheese mainly through the macropores and secondly through the protein matrix. Most of the ^{131}I must be in an ionic than non-ionic form and this fact justifies the larger amounts of ^{131}I penetrating than removed. Most probably the ionic ^{131}I penetrates in chemical covalent bonds with organic substances in cheese protein matrix [13]. The presence of salt in water is found to decrease the penetration rate of iodine in cheese, compared to clear water (Table 3.36).

Actually, the cluster of nine water molecules that surrounds each of the sodium ions in the hydration zone together with the ^{131}I, stimulate penetration at much lower rate. Also, the tendency of ^{131}I to participate in chemical covalent bonding with proteins is decreased because of possible salting-out effects [115].

Therefore, NaCl or KCl enhances the removal of ^{131}I from contaminated cheese to brine and retards the penetration of ^{131}I from contaminated brine to uncontaminated feta cheese.

Conclusions

The mechanism of ^{131}I transport in feta-type cheese is a complex process and it cannot be explained only by a constant diffusion coefficient. It will be easier to understand the removal or penetration of ^{131}I by taking in consideration the macroscopic transport of ^{131}I along the cheese macropores and its interaction with the protein matrix. During this process, some of the interacting ^{131}I molecules are trapped inside the cheese protein lattice. The amount of trapped iodine depends on the state of cheese proteins and the concentration of neutral salt [71]. Therefore, the value of the diffusion coefficient found is time dependent. It can be assumed, however, that for less than five hours immersion time, the transport of ^{131}I along the cheese macropores is regulated by a constant diffusion coefficient which varies from 5.5×10^{-5} cm^2 s^{-1} to 6.4×10^{-5} cm^2 s^{-1}.

3.3.3 Study of ^{131}I Impact Over Lactic-Acid Micro-Flora of the Yogurt

Tables 3.37, 3.38, 3.39, 3.40 and 3.41 present the results of the studies carried out, concerning ^{131}I impact over the lactic-acid fermentation. The chemical composition of the sheep and cow milk used for experimental production of yogurt is presented in Table 3.37 [186–189, 223].

Table 3.37 Chemical composition of milk used

Type (g/100 g sample)	Protein	Fat	Lactose	Ash	Total solids
Cow	3.8[a]	3.9	4.8	0.8	13.3
Sheep	5.3	6.0	4.8	0.8	16.9

[a] Each value is a mean of four replicate determinations

Table 3.38 Chemical composition of set and strained yogurt made from cow and sheep milk

Sample (g/100 g)	Types					
	Cow milk			Sheep milk		
	Set	Strained	Whey	Set	Strained	Whey
Protein	4.2	10.2	0.2	5.9	12.7	0.3
Fat	4.4	11.0	0.0	6.7	14.8	0.1
Lactose	4.3	4.1	4.1	4.1	3.9	3.8
Ash	0.9	1.6	0.5	0.9	1.4	0.6
Lactic acid	1.0	2.0	0.7	1.2	2.1	0.9
Total solids	14.8	28.9	5.5	18.8	34.9	5.7

The amounts used vary from 6,000 to 12,000 Bq/kg of milk. The traditional–the so-called set and strained–the so-called concentrated cow milk have composition, similar to that of the sheep milk (Table 3.38).

The yield of set yogurt from cow milk and sheep milk was approximately 90%. The yield of concentrated yogurt from cow and sheep milk was 42 and 46% respectively. The possible reasons for different results obtained for the concentrated yogurt may be due to differences in milk composition for both species and the duration of draining whey from set yogurt. It seems that the amount of radioactivity did not affect the yield and the chemical composition of two types of yogurt compared to control samples. The microbial counts of set or concentrated yogurt are presented in Tables 3.39 and 3.40. In general, any difference obtained in microbial counts of set or concentrated yogurt can be attributed to the differences in composition of milks since the same start microbial culture (yeast) was used for production of both types of yogurt after pasteurization of raw milk.

Percentages of the lactic-acid bacteria (LAB), which survive in both types of yogurt and whey, are presented in Table 3.41.

It is quite clear from Tables 3.39, 3.40 and 3.41 that the isotope ^{131}I inactivates a great percentage of lactic acid bacteria. This is more obvious in the case of strained yogurt where there is more time and quantity to ^{131}I available to react with the bacteria. It also seems that in the case of strained yogurt the amount of lactic acid bacteria available affects the resistance to lactic acid microbial species used in the production of yogurt. *S. thermophilus* is more sensitive to ^{131}I compared to *L. bulgaricus*. This difference in sensitivity to ^{131}I between the two microorganisms may cause problems in structure, taste and the quality of yogurt, due also to proteolysis, which is faster in the presence of *L. bulgaricus*.

Table 3.39 Viable cell counts of *L. bulgaricus* and *S. thermophilus* in yogurt and whey from cow milk

Product	*L. bulgaricus*		*S. thermophilus*	
	CFU × 108/ml sample		CFU × 108/ml sample	
	Control	Radioactive product	Control	Radioactive product
Set Yogurt	705.2[a] (23.2)	368.8 (15.6)	598.0 (19.8)	193.4 (12.4)
Strained Yogurt	701.3 (21.5)	167.1 (11.7)	338.3 (14.9)	99.3 (5.6)
Whey	3.1 (0.2)	1.4 (0.1)	3.0 (0.2)	1.1 (0.1)

[a] Each value is a mean of four replicate determinations (standard error of the mean in parentheses)

Table 3.40 Viable cell *L. bulgaricus* and *S. thermophilus* in yogurt and in the serum from sheep milk

Product	*L. bulgaricus*		*S. thermophilus*	
	CFU × 108/ml sample		CFU × 108/ml sample	
	Control	Radioactive product	Control	Radioactive product
Set Yogurt	719.2[a] (23.2)	375 (15.9)	591.1 (19.8)	181 (12.1)
Strained Yogurt	718.1 (25.9)	175.2 (15.7)	310.4 (15.5)	111 (7.9)
Whey	3.2 (0.2)	1.5 (0.1)	3.1 (0.2)	1.2 (0.1)

[a] Each value is a mean of four replicate determinations (standard error of the mean in parentheses)

Table 3.41 Lactic-acid bacteria (LAB) surviving in both types of yogurt and whey

Type	Product	%[a]	
		L. bulgaricus	*S. thermophilus*
Cow	"Set" yogurt	52.2	32.3
	"Strained" yogurt	23.8	29.4
	Whey	45.2	36.7
Sheep	"Set" yogurt	52.1	30.6
	"Strained" yogurt	24.4	35.8
	Whey	46.9	38.7

[a] Percentage comparing radioactivity to control sample

The decreased viability of *L. bulgaricus* and *S. Thermophilus* in the presence of ^{131}I may be due to the ionization of a portion of the water molecules with the formation of highly reactive hydrogen and hydroxyl radicals, which inactivates the enzyme system in both types of yogurt and their microbial content [111, 144]. It is believed that the indirect effects of ionizing radiations are apparently largely responsible for killing microorganisms [45].

Conclusions

The results demonstrate that the radionuclide ^{131}I decreases viability of *L. bulgaricus* and *S. thermophilus* by 45% in set and strained yogurt prepared from artificially contaminated cow or sheep milk. The presence of ^{131}I had no or minimal effect on the gross composition and percent yield of yogurt prepared from both kinds of milk. Biological testing of yogurt types is required to help to elucidate the possible nutritional differences that exist in radioactive contamination. The preservation of lactic acid micro-flora in yogurt prepared from radioactively contaminated milk requires regulation of its production parameters.

3.3.4 Decontamination of Milk from the Radioactive Isotopes ^{131}I and $^{134,137}Cs$

Milk is one of the basic human nutriciants and this is the reason why have been made in the past enough efforts for its decontamination from significant radionuclides resulting as a consequent product of nuclear explosion or leakages from nuclear installations. The main attention of decontamination has been concentrated on the removal of iodine [131], cesium and strontium (when existing in an amount great enough), presenting, from a radiobiological aspect, the greatest interest and from the other side they constitute the greatest percentage of radioactivity in milk. For the decontamination are used ion exchange resins [47], which are cationic for the removal of cesium and strontium, and anionic for iodine (Table 3.42). The decontamination is completed when the final pro-duct has a final radioactivity lower than the limits defined by the Radiation Protection authorities of the country [197, 223]. For instance, after the Chernobyl accident the total maximum radioactivity in milk should not exceed 500 Becquerels (Bq):

$$1^{131}\mathrm{I} + 0,25^{137}\mathrm{Cs} + 0,33^{134}\mathrm{Cs} < 500\ \mathrm{Bq}$$

The technological line of the flow of processes has been found. It includes the necessary and auxiliary products.

The description of the flow diagram for processes in-cludes the main (thick black arrows) and the possible recycles which are required to face special cases or situations, is shown in Fig. 3.29. The technological scheme starts with milk, which is stored in a special tank (at pH 6.84 and temperature 2°C). By regulation of the temperature with the heat-exchange units, the first step of the process is

Table 3.42 Characteristics of the ion-exchanging resins, used in the ion-exchanging columns

Property	Resin	
	Amberlite IRA 120	Amberlite IRA 904
Character	Cationic	Anionic
Active group	$-SO_3^{2-}$	$-N-(CH_3)_3^+$
Form	H	Cl
Phenomenical density	0.84 g/cm^3	0.67 g/cm^3
Real density	1.27 g/cm^3	1.08 g/cm^3
Volume density	850 g/l	670 g/l
Active grain metric	0.45–0.60 mm	0.40–0.50 mm
Humidity	44–48%	56–62%
pH range	0–14	0–14
Maximal temperature	120°C	77°C
Total alternative capability (meq/ml)	1.9	0.7
Swelling after the total conversion	7%	5%

accomplished, i.e. ^{131}I (anions) removal in the anionic ion-exchanger column. When passing through the anionic ion-exchanger column, the milk's temperature is 13°C. The columns consisting of anionic and cationic ion-exchangers have compulsory been previously sterilized with hot steam. They are also expected to be regenerated. After the milk passes through the anionic ion-exchanging column, its radioactivity starts being controlled. Because of the high presence of iodine, past the acceptable limits, milk is driven several times through the anionic ion-exchanging column until the acceptable values of iodine are obtained.

After being purified from ^{131}I, the milk is treated by a solution of citric acid, so that it obtains pH 5.30. The temperature is expected to be 9°C, which is obtained with the help of a cooling device. Then filtration is the next stage of the process. The filtration purpose is keeping the possible structures or colloids, which may hinder the normal functioning of the cation ion-exchanging columns. The milk temperature at the exit of the cationic columns shall be about 13°C.

The control of cesium concentration (^{134}Cs and ^{137}Cs) in milk is a regulating factor, which leads milk to be recycled and to pass through the cationic ion-exchanging columns again or for the final processing phase of neutralization of the milk. Neutralization is carried out by KOH solution, which flows in from the relevant tank. The pH value of the milk is restored to 6.76.

The following filtration step is necessary for the removal of all unwanted solid substances from the final product. After that the process of reprocessing continues. Some of the processes are annulled when unnecessary (for example passing through the anionic column in case there is no iodine). Eventual errors (wrong acidification or neutralization, errors in mounting devices, etc.) are eliminated.

Conclusions

There has been studied a possibility of milk purification from radioisotopes (^{131}I and $^{134,\ 137}$Cs) by ion-exchanging resins. On the ground of the results

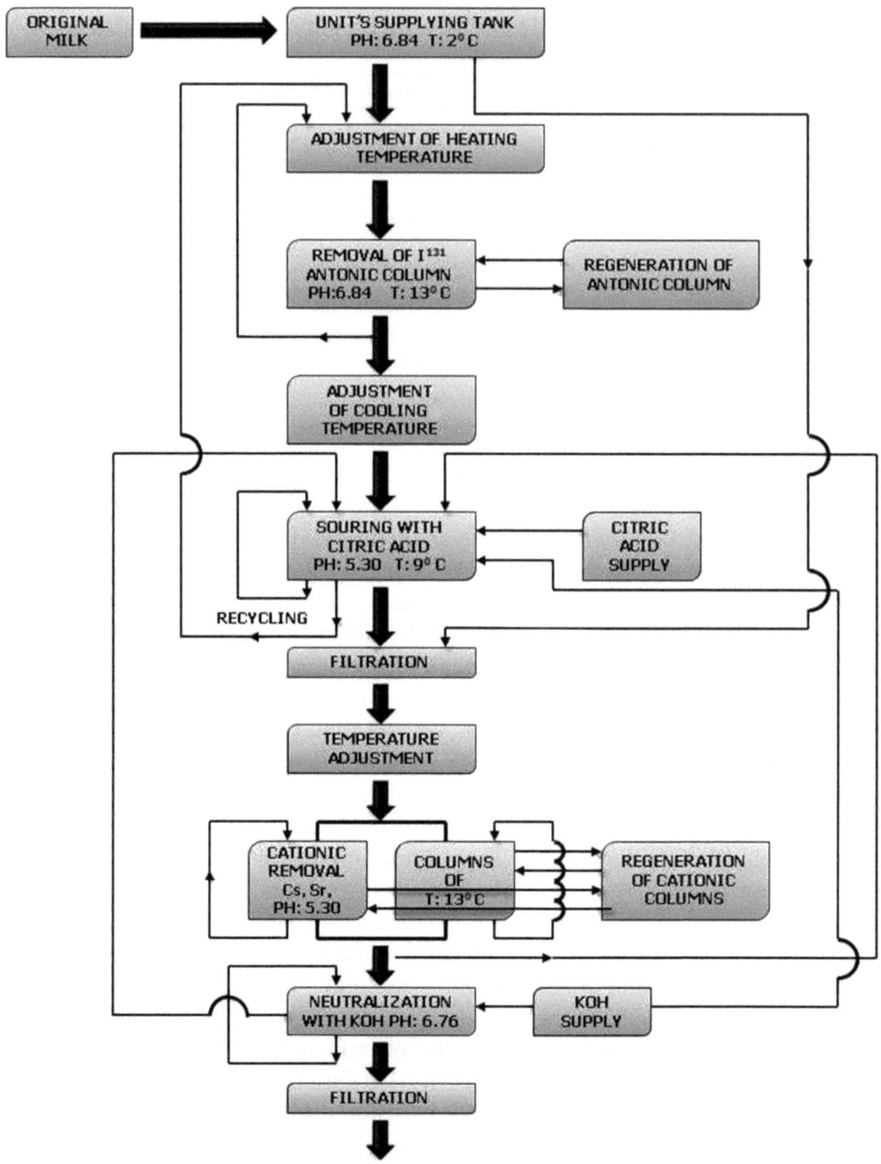

Fig. 3.29 Flow diagram from the decontamination of milk from the radioisotopes iodium-131 and cesium-134, 137

obtained, the elements of a model of technological line for milk purification by ion-exchanging resins have been determined.

3.3.5 The Preservation and Maturing of Diary Products by using an Irradiation Process

According to recent estimations of the World Health Organisation (WHO), nearly 20% of the world's food production is destroyed because of deterioration. This can be prevented either with preservatives or with physical and biological methods, such as drying, cooling, freezing, etc.

This method, mentioned above, has been successfully used to food deterioration. Although irradiation has been researched for years it has only recently been used to commercially preserve foods. The foodstuff is exposed to high energy radiation that decreases preservations time and improves both quality and safety [200, 202, 207].

The γ-rays, which were used for the present irradiation, are emitted from the radioactive material Cobalt-60 (^{60}Co). This is an artificial radioisotope made by bombardment of the stable isotope Cobalt-59 (^{59}Co) with neutrons in a nuclear reactor. The radioisotope ^{60}Co emits electrons and γ-rays that have energy of equivalent 1.17 and 1.33 MeV, respectively. The average of γ-radiations of ^{60}Co is 1.25 MeV [203]. γ- and X-rays as well as electrons, interact with the matter through many phenomena, relatively to their energy. γ-rays energy is absorbed progressively from the matter, while the electrons are retarded and stop in a significant depth in the material. In accord, γ-rays are usually used in the processing of food that has high density and the electron beam only in food with low density and small thickness [204]. The exposure time of the foodstuff, which is going to be processed, depends on the power, the productivity and the geometry of the installation [201].

The irradiation process changes the molecular structure of the foodstuff and a number of molecules return to their ground state, which others obtain a new structure. The new molecules, which are produced from the radiation effect, are called "actinolysis products" and they are in a very small amount because the time of exposure of the foodstuff to the radiation is relatively limited. Therefore, with 1 Mrad per kg of foodstuff are formed less than 20 mg products of actinolysis, which is composed of 80% H_2O, 6.6% lipids and 6.7% proteins. There is a deterioration of carbohydrates which give to the foodstuff the typical flavor and its nutritional value. Furthermore, 20 years of research on guinea pigs fed with foodstuff or with products of actinolysis concluded that there were no potential toxic products or products which can cause alternations or cancer [205].

Although the production of actinolysis products, whose amount is too small to destroy the neutrality of foodstuff, the radiation can have other consequences on some molecules. In foodstuffs, the possible DNA change has no consequences on cells which have already completed their biological cycle. On the contrary, the DNA change is important, when it happens on cells which have the capability of reproduction, because it can lead them to alteration, to a loss of their capability of reproduction or to a retardation of the operations. The food is irradiated beyond the DNA and consequently this brings changes to the DNA of the microorganisms and parasites found in food. As a result, some of them which can have toxic effects are killed, while the insects, due to their sexual chromosomes attacked, are becoming sterile. The parasites are totally vanished since their cells are destroyed [246, 247].

Generally, the results of the radiation depend on the amount of energy absorbed by the foodstuff and the sensitivity is presented by the "stroke" atoms. It has been calculated that for food preservation; the required radiation doses vary from 0.1 to 6 Mrad.

In this study samples of kasseri (type of pasta filata cheese) have been irradiated and also unprocessed cheese (baski), aiming the preservation capability of the first and faster mature of the second [198, 250]. The experiment lasted four months and three series of irradiation have been taken place. The irradiation of the samples was done in a special apparatus (bomb) of cobalt (^{60}Co), which emits γ-rays of an average energy of 1.25 MeV [208]. The low-dose rhythm of the source and its simultaneously use for other activities, were disadvantages for the present work, since every cycle of irradiation lasted many days. The rhythm of dose in the center of irradiation chamber was about 7.5 rad s^{-1}, which means that a sample to be irradiated with a dose of I Mard, had to stay in the chamber about 35 h. The irradiation cahmber had the shape of a cylinder with a diameter of 8 cm and a height of 13 cm.

First cycle of irradiation

Kaseri and baski from the market were placed immediately in the refrigerator. After a few days, a part of kasseri was placed in the freezer and the following samples were prepared:

M	from baski which was preserver in the refrigerator
K	from kaseri which was preserver in the refrigerator
Y	from kaseri of the dairy
S	from kaseri which was covered offhand with nylon

These four series of samples (Table 3.43) were irradiated with doses from 1 to 150 Krad (1.5 kGy). Due to the low dose rhythm of the used source the big number of samples; totally long irradiation time was needed. The samples No. 5 and 7, (30 and 70 Krad or 0.3 and 0.7 kGy) were irradiated respectively.

The size of the first samples, where the irradiation cycle was completed in a suitable period of time, was unfortunately too small. Therefore, it was not possible to do reliable biological and biochemical analysis.

After the end of the irradiation, the samples Y and K were separated into two parts. One part was places in the refrigerator while the other one was kept at room temperature (20–21°C). From the first irradiation cycle, due to the small doses only optical conclusions could be derived.

Second cycle of irradiation

Two groups of samples were prepared and stored in the refrigerator: M from baski and Y from kasseri (Table 3.44). Irradiation doses were much higher than those of the first cycle. Table 3.45 shows the microscopic analysis of the samples.

Table 3.43 First cycle of irradiations

Sample				Radiation dose (Krad)
M_1	K_1	Y_1	S_1	1
M_2	K_2	Y_2	S_2	5
M_3	K_3	Y_3	S_3	10
M_4	K_4	Y_4	S_4	20
M_5	K_5	Y_5	S_5	30
M_6	K_6	Y_6	S_6	40
M_7	K_7	Y_7	S_7	50
M_8	K_8	Y_8	S_8	70
M_9	K_9	Y_9	S_9	100
M_{10}	K_{10}	Y_{10}	S_{10}	150

Table 3.44 Second cycle of irradiation

Sample		Radiation dose (Krad)
Y_{20}	M_{20}	170
Y_{35}	M_{35}	610
Y_{50}	M_{50}	760

Table 3.45 The irradiation doses with the corresponding microscopic analysis

Sample	Enterococci	Lactobacilli	Z/M	Dose (Krad)
Y_0	700	4.10^6	18×10^3	0
Y_{20}	50	$<10^5$	12×10^3	170
Y_{35}	50	$<10^5$	7×10^3	610
Y_{50}	0	$<10^5$	12×10^3	760

Third cycle of irradiation

In this cycle decided the size and packing forms o the irradiated kasseri samples, according to the market standards. The irradiation time and doses were high. The results of these samples and control samples are shown in Table 3.46. The microscopic analysis of the samples of this cycle is shown in Table 3.45.

A sample (N_1) was irradiated (640 Krad) and compared with a control sample (N_2), to see if maturing can be accomplished by irradiation. The maturing factor (M.F.) was calculated from the microscopic–chemical analysis [206]:

$$\text{M.F.} = \frac{\text{Water-solube nitrogen }\%}{\text{Total nitrogen}}$$

with the following obtained results:

$$(\text{M.F.})\,N_1 = 0.784/3.89 = 20.15\%$$

Table 3.46 The irradiation doses on samples and control samples

Sample	Dose (Mrad)	Control sample
Z_{15}	1.1	Z_{40}
Z_6	2.7	Z_4
Z_2	3.0	Z_{30}
Z_5	4.8	Z_3
Z_{12}	5.8	Z_7

Table 3.47 The irradiation doses with the corresponding microscopic analysis

Sample	Dore (Mrad)	Enterococci g^{-1}	Coliforms g^{-1}	Lactobacilli g^{-1}	Yeast and fungi g^{-1}
Z_{40}		150	0	17×10^6	12.10^3
Z_{30}		50	0	55×10^6	3×10^3
Z_3		100	0	18×10^6	10^4
Z_7		125	0	25×10^6	12×10^3
Z_{15}	1.1	0	0	2×10^4	<1000
Z_2	3.0	0	0	2×10^4	1000
Z_5	4.8	0	0	$<10^4$	1000
Z_{12}	5.8	0	0	$<10^4$	<1000

and

$$(\text{M.F.})\text{N}_2 = 0.532/3.73 = 14.25\%$$

Irradiation accelerated maturing about 5.9%. The rest of the irradiated and control samples were examined and their microscopic analysis results are shown in Table 3.47.

For the killing of all organisms, that causes food deterioration, doses of 5.6 Mrad are needed. All the proteins and amino acids sustain without significant reduce of their nutritional value, when the radiation levels are regulated to levels suitable for industrial processing. Basically, radiation causes chemical changes at fats, due to heating and oxidation reactions and of a great number of produced compounds. The basic reactions that take place are oxidations, polymerization, decarboxylation and dehydration.

Regarding the induction of the maximum radioactivity in the irradiated foodstuff the following equation was used [45]:

$$Q = \frac{30\ \text{Den}}{T}$$

T = half-life of radioisotope
Q = radioactivity in pCi
D = dose in Mrad
E = energy in MeV
N = decay constant

The obtained result was practical zero. Microorganisms causing diseases (annihilations of *enteroccoci* and a relative reduction yeast) are destroyed at irradiation doses of 0.5–0.7 Mrad (Table 3.45). These are some losses either in nutritional value or useful microorganism caused by irradiation of because the time is identical for destruction of desirable microbes connected with cheese mature *(lactobacilli)*. In cases of drastic reduction (around 1,000 g^{-1}) of yeast and of course annihilation of *enteroccoci,* the destruction of *lactobacilli* is almost rapid (around 10^4 g^{-1} out of normal 25×10^6 g^{-1}), (Table 3.47). This fact can cause deterioration of organoleptic characteristics of cheese or some disturbances during maturing after the exportation from the factory. An alteration of the yellow color of cheeses has been observed with the irradiation doses greater than 2 Mrad, while the taste was not altered significantly. For kasseri the irradiation range was among 400–650 Krad (4–6.5 kGy).

Conclusions

The destruction of unwanted microorganisms was a satisfactory one and the decrease of wanted microorganisms was relative small. Of course, all the subject aspects, have not been examined especially regarding the development of unwanted by-products such as toxic ones. Finally it has been shown clearly that from the present work using a dose of 640 Krad during the third cycle of irradiation increased the maturing factor up to 6% (microscopic analysis of cheese sample N_1 and control sample N_2). This means practically that maturing period could be shortened from 15 days (naturally maturation) to 11 days by irradiating kasseri at a temperature of 11°C.

3.3.6 Investigation of Natural Radioactivity Concentration in Building Materials for Interior and Exterior Adornments in Modern Greek Style Constructions

The aim of the present work is to determine the specific activity concentrations (Bq kgr^{-1}) of ^{238}U, ^{232}Th and ^{40}K presented in building materials used for construction and decorative purposes inside the houses [155, 199, 227, 236]. These are cement, cement brick, asbestos, red bricks, sand, solid and soft rocks, marbles, limestone, granite, iron, aluminium, copper and brass collected from various places in Greece. Direct gamma counting spectroscopic method is used for acquiring the values. These values are used in order to estimate indirectly the annual radiation dose in mSv that humans may acquire and thus to assess the possible radiological hazard to humans due this natural irradiation [209].

Finally, the results are evaluated by comparing them with international standards of permissible doses. However, it should be mentioned that the natural radiation dose received by individuals depends on factors such as rates of ventilation, pattern of airflow, time, etc. [248].

A total of 31 building materials are crushed into fine granules. Each of them was homogenized and air-dried. Approximately 50 ml of each of the samples was filled into small cylindrical plastic containers (height 2 cm, diameter 7 cm). Their respective net weights were measured and recorded with a high sensitive balance. Then these samples were sealed and left airtight for 4 weeks to allow for radium and its short progeny to be in radioactive equilibrium.

The gamma-ray spectra of the samples were collected using a high resolution HPGe detector (Eurisys High Purity Germanium Coaxial P-Type Detector) with a photopeak relative efficiency of 20% and energy resolution (FWHM) 1.80 keV for the 1332 keV energy of ^{60}Co [227].

Through a pre-amplifier, the detector was connected to the amplifier, which was positioned inside the PC main box. An ADC of 8192 channels was assembled to the system and the spectroscopic measurements and analysis was performed via the Oxfordwin Assayer software packed into the PC of the Laboratory [228].

In order to reduce gamma-ray background the detector was maintained in a vertical position and shielded by a cubic type iron shield of 32 cm height and 21 cm depth, with a front moving door. The external iron shield contained two inner cubes made of 4 cm thickness of lead and 0.5 cm thickness of copper.

Absolute efficiency calibration of the detector was carried out by a 50 ml standard solution of ^{152}Eu inside a glass container (for energies from 122 to 1530 keV) using the same geometry as the measured samples. Corrections for self-absorption effects in the samples were found to be insignificant for gamma rays with energies greater than 400 keV [210].

The activity of ^{238}U was estimated from the 609.3 keV gamma transition energy of ^{214}Bi (41.2% possibility). Also the activity of ^{232}Th was estimated from the 583.1 keV gamma transition energy of ^{208}Tl (86% possibility) while ^{40}K activity was determined using the 1460 keV gamma ray (10.7% possibility) [114]. However other gamma transition energies may be selected in limited cases where the total error of the measurement was inevitably very low. Background activities for the three selected energies were subtracted from the sample's readings in order to assess the true activities of each isotope.

The samples were placed coaxial 10 cm from the surface of the detector and the counting time for each sample were 20 h.

The activity concentration of each isotope was calculated using the following equation:

$$A_s = C_s / \varepsilon P_r M_s (\mathrm{Bq\ kg^{-1}}) \tag{3.5}$$

Where,

A_s is activity concentration of the radionuclide (Bq kg^{-1})
C_s is counting rate of the gamma ray (counts per second)
ε is detector efficiency of the specific gamma energy
P_r is absolute transition probability of gamma decay
M_s is mass of the sample (kg)

Table 3.48 Specific activities of ^{238}U, ^{232}Th and ^{40}K isotopes present in various indoor building materials

Samples	Specific activities (Bq kg^{-1})			$^{226}Ra_{eq}$ (Bq kg^{-1})	Maximum permissible dose
	^{238}U	^{232}Th	^{40}K		
Red clay brick	54 ± 24	61 ± 22	507 ± 63	180 ± 60	Below
Asbestos	19 ± 22	7 ± 36	0 ± 91	27 ± 80	Below
Cement brick	31 ± 36	0 ± 11	172 ± 80	44 ± 58	Below
Cement	29 ± 25	3 ± 23	361 ± 55	61 ± 62	Below
Sand	17 ± 19	4 ± 14	0 ± 63	23 ± 44	Below

The ^{226} Ra_{eq} values are also presented

Radium equivalent activity (Ra_{eq}) of each building material was calculated according to Beretka and Mathew [211] as follows:

$$Ra_{eq}(\mathrm{Bq\,kg^{-1}}) = A_{Ra} + (A_{Th}1.43) + (A_K\,0.077) \quad (3.6)$$

where,

A_{Ra}, A_{Th} and A_K are the activity concentrations of uranium (^{238}U), thorium (^{232}Th) and potassium (^{40}K) in Bq kg^{-1}.

This equation is based on the estimation that 370 Bq kg^{-1} of ^{238}U, 259 Bq kg^{-1} of ^{232}Th and 4810 Bq kg^{-1} of ^{40}K produce the same gamma ray dosage.

Also the Ra_{eq} value of 370 Bq kg^{-1} is equivalent to the annual dose equivalent of 1.5 mSv per year, which we assumed to be the maximum permissible dose to humans from their exposure to natural radiation from building materials in one year [212].

Tables 3.48, 3.49, 3.50 and 3.51 shows the gamma ray activity concentrations of uranium (^{238}U), thorium (^{232}Th) and potassium (^{40}K) in Bq kg^{-1}, measured for a total of 31 building materials used for various purposes in most of the Greek home constructions such as walls, roofs, floors, kitchen benches and fireplaces. In these tables, radium equivalent activities of each material were estimated using the Eq. (3.6). The last column presents a statement of whether the radium equivalent activities of each building material are above, below or near the Ra_{eq} value of 370 Bq kg^{-1}.

As shown in Table 3.48, the specific activities of uranium (^{238}U) for all samples, which are categorized as the basic components of the Greek construction, ranged from 17 to 54 Bq kg^{-1}. The lowest value was found in the sand, while the highest value was observed in the red clay brick. Thorium (^{232}Th) specific activities ranged from zero to 61 Bq kg^{-1} for red clay brick. Finally, potassium (^{40}K) specific activities ranged from 0 to 507 Bq kg^{-1} for red clay brick. For red-clay brick the values obtained for uranium, thorium and potassium well exceed the World and Greek average activities for soil (almost two times higher). Also the activity concentrations for uranium and potassium for the cement are very close to the World average activities for soil. Radium equivalent ($^{226}Ra_{eq}$) activities of the red-clay brick exceed the World and Greek average radium equivalent values.

Table 3.49 Specific activities of ^{238}U, ^{232}Th and ^{40}K isotopes present in various building materials used primary *for decorative purposes*

Samples	Activity concentration (Bq kg^{-1})			$^{226}Ra_{eq}$ (Bq kg^{-1})	Maximum permissible dose
	^{238}U	^{232}Th	^{40}K		
Solid rock of Pilio	121 ± 25	148 ± 24	2024 ± 103	488 ± 67	Above
Solid rock of Caristos	57 ± 41	10 ± 25	302 ± 94	95 ± 84	Below
Soft rock of Hios[a]	42 ± 23	36 ± 38	0 ± 91	93 ± 84	Below
Schist-rock of Thessaloniki	42 ± 24	84 ± 23	1490 ± 97	277 ± 64	Below
Soft rock of Hios[b]	129 ± 23	130 ± 37	153 ± 93	327 ± 83	Near
Granite of Thrace	92 ± 38	708 ± 36	2105 ± 153	1267 ± 101	Above
Limestone of Thasos	43 ± 41	0 ± 21	0 ± 85	43 ± 78	Below
Redish stone of Thasos	24 ± 21	8 ± 20	216 ± 89	52 ± 56	Below
Limestone of Crete	61 ± 22	7 ± 21	0 ± 85	71 ± 59	Below
Soft rock of Rodos	55 ± 26	59 ± 24	193 ± 102	154 ± 68	Below

[a] and [b] are two different samples of different rocks from the Island Hios

In Table 3.49 the specific activities of uranium (^{238}U) for all samples, which are categorized as complementary materials for interior and exterior adornments in the Greek construction, ranged from 42 to 121 Bq kg^{-1}. The same for the Thorium (^{232}Th) specific activities, which was ranged from 0 to 708 Bq kg^{-1} and for potassium (^{40}K) specific activities which was ranged from 0 to 2024 Bq kg^{-1}. It is obvious that uranium (^{238}U) activities for all these materials exceed the World and Greek average activities for soil (until three times higher for the solid rock of Pilio). However, thorium (^{232}Th) and potassium (^{40}K) activities are found to be from very low (zero) to very high (2024 Bq kg^{-1}) and thus cannot give a safe conclusion by comparing them with the activities of the World and Greek soil. Radium equivalent ($^{226}Ra_{eq}$) activities of the majority of the special materials are shown to exceed the World and Greek average radium equivalent values for the soil. As the maximum radiation dose to be emitted by building materials is 1.5 mSv per year, which corresponds to the maximum radium equivalent activity of 370 Bq kg^{-1}, we observe that all the main building materials from Table 3.49 are below this limit. On the other hand two special materials from Table 3.49 (solid rock of Pilio and Granite of Thrace) are well above this limit and one special material (Soft rock of Hios^2) is near this limit value.

As shown in Table 3.50 the specific activities of uranium (^{238}U) for all samples, which are used for interior Greek house construction, range from 56 to 140 Bq kg^{-1}. The same for the Thorium (^{232}Th) specific activities, which range from 27 to 132 Bq kg^{-1} and for potassium (^{40}K) specific activities, which range from 378 to 1978 Bq kg^{-1}. As the maximum radiation dose to be emitted by building materials is 1.5 mSv per year, which corresponds to the maximum radium equivalent activity of 370 Bq kg^{-1}, we observe that two indoor building materials (Solid rock of Volos and Marble of Dry River Dramas) are well above this limit and two of them (Redish rock of Livadia and Marble of Pigon Dramas) are near this limit value if we consider the relatively high inherent value error.

Table 3.50 Specific activities of ^{238}U, ^{232}Th and ^{40}K isotopes present in various indoor building materials used primary *for decorative purposes*

Samples	Activity concentration (Bq kg^{-1})			$^{226}Ra_{eq}$ (Bq kg^{-1})	Maximum permissible dose
	^{238}U	^{232}Th	^{40}K		
Solid rock of Volos	113 ± 21	132 ± 22	1978 ± 97	454 ± 60	Above
Purple rock of Ioannina	72 ± 51	27 ± 12	978 ± 67	186 ± 73	Below
Redish rock of Livadia	87 ± 62	72 ± 49	1240 ± 94	285 ± 139	Below
Schist-rock of Kavala	59 ± 32	64 ± 29	1365 ± 55	256 ± 78	Below
Marble of dry river Dramas	140 ± 44	126 ± 37	894 ± 63	389 ± 102	Above
Marble of Pigon Dramas	97 ± 51	86 ± 39	754 ± 62	278 ± 112	Below
Marble of Volaka	63 ± 27	48 ± 39	654 ± 52	182 ± 87	Below
Marble of Kozani	56 ± 32	58 ± 28	421 ± 41	171 ± 75	Below
Marble of Nestos	62 ± 26	44 ± 22	378 ± 40	154 ± 61	Below
Marble of Veria	88 ± 36	65 ± 24	556 ± 34	224 ± 73	Below

Table 3.51 Specific activities of ^{238}U, ^{232}Th and ^{40}K isotopes present in metallic materials

Samples	Activity concentration (Bq kg^{-1})			$^{226}Ra_{eq}$ (Bq kg^{-1})	Maximum permissible dose (370 Bq kg^{-1})
	^{238}U	^{232}Th	^{40}K		
Iron 1	79 ± 32	74 ± 21	990 ± 74	261 ± 68	Below
Iron 2	62 ± 29	51 ± 22	730 ± 56	191 ± 65	Below
Aluminium 1	76 ± 32	61 ± 33	1026 ± 71	242 ± 85	Below
Copper	112 ± 42	87 ± 36	1239 ± 112	332 ± 102	Near
Aluminium 2	92 ± 33	71 ± 27	982 ± 91	269 ± 79	Below
Brass	97 ± 30	65 ± 24	987 ± 91	266 ± 71	Below

The 1 and 2 index at the iron and aluminium samples indicates that they are from different brands

As shown in Table 3.51 the specific activities of uranium (^{238}U) for all samples, which are used for interior Greek house construction, range from 76 to 112 Bq kg^{-1}. The same for the Thorium (^{232}Th) specific activities, which range from 51 to 87 Bq kg^{-1} and for potassium (^{40}K) specific activities, which range from 730 to 1239 Bq kg^{-1}. As the maximum radiation dose to be emitted by building materials is 1.5 mSv per year, which corresponds to the maximum radium equivalent activity of 370 Bq kg^{-1}, we observe that only the copper sample is near this limit (Table 3.52).

Table 3.53 compares radium equivalent activities for selected typical main building materials from other countries. The content of the same material can vary from brand to brand and the total activities depend on the radioactivities of the components in the different materials. As shown in Table 3.53 the values of radium equivalent obtained for these materials are found to be the lowest compared to that of other countries.

Table 3.52 The evaluated *average* activities of ^{238}U, ^{232}Th, ^{40}K and $^{226}Ra_{eq}$ isotopes present in soil

	Activity concentration (Bq kg^{-1})			$^{226}Ra_{eq}$ (Bq kg^{-1})
	^{238}U	^{232}Th	^{40}K	
World average activity of soil [211]	25	28	370	89
Greek average activity of soil [212]	45	47	460	147

Table 3.53 Comparison of the mean radium equivalent activity concentrations (Bq kg^{1}) of various building materials of several studies

Materials	Greek (2001)	Greek (2000)	Malaysia (1996)	Zambia (1995)	Australia (1985)	Germany (1981)
Red clay brick	54 ± 24	183 ± 17	895 ± 107	180	833	640
Asbestos	17 ± 22	23 ± 2	1131 ± 157	61	–	–
Cement	29 ± 25	53 ± 4	188 ± 27	79	115	70
Sand	17 ± 19	10 ± 1	136 ± 33	135	70	59

It is important here to notice that the biological effects to humans caused by a natural radiation source such as our samples are divided into stochastic and non-stochastic effects. The difference between them is that non-stochastic effects have a threshold value and this value is important not to exceed instead of the stochastic effects that there is no threshold and the probability of occurrence depends on dose linearity down to zero dose but the severity of the effect is independent of the dose [213].

The threshold value here is 1.5 mSv y^{-1} and in order to prevent by exceeded it is to keep away from natural sources which seems to exceed this value.

However there are some radiation protection rules [214] that we can follow in order to minimize radiation dose intake from that kind of indoor materials. These are:

1. to increase the distance between the individual and the material (as far as possible) utilizing the inverse square law of the radiation
2. to minimize the time that and individual is in close distance with the material
3. to use protective barriers if this is possible

Finally, it has to be considered that different radiation sources give an additive whole body radiation dose to an individual that is in distance with them [215]. This means that building materials used for housing construction must *not* give a total value of natural radiation dose higher than the maximum permissible value of 1.5 mSv y^{-1}.

Conclusions

The present results have shown that most of the building materials used in Greek construction meet the criterion limit of radiation dose to humans for building materials (1.5 mSv per year). A limited number (three out of ten) of decorative special type materials collected from different places in the whole Greece have shown an increased dose.

In addition, the majority of the main Greek building materials show lower radium equivalent activities than the World and Greek average activity of soil with an exception of the red clay brick. In contrast, the majority of the Greek building materials used for interior and exterior adornments show much higher radium equivalent activities comparing to the World and Greek average activity of soil.

However, comparing these results with similar data of other countries, the Greek samples have shown the lowest radium equivalent activities [216–218].

In the future the study should concentrate to a more systematic investigation of the natural radioactivity contained in building materials for interior and exterior adornments from different brand and from different regions in Greece as these materials are found from this study to be the most contaminated with natural radioactive radionuclides.

3.3.7 The Marine Radioactivity

The marine environment has received man made radio nuclides from many sources: the tests of nuclear weapons in the atmosphere, the controlled discharges of low level liquid effluents from the nuclear power industry, fallout arising from accidents such as the USSR Chernobyl reactor accident in 1986, the transit 5BN-3 satellite accident and other minor sources such as laboratories at universities and research centers, phosphate ore processing, etc. These sources are defined as the impact sources of a certain marine environment in case. The impact source of radioactive pollution in the Northeastern Mediterranean has been defined as the mouth of the Dardanelles Straits, in the North Aegean Sea. As the Aegean Sea is adjacent in its northern part to the Black Sea through the Straits of the Dardanelles, the Black Sea water mass enters the Aegean Sea via surface current circulation during the water exchange procedure between the two basins. The ^{137}Cs levels detected until April 1986 were due to worldwide fallout and varied between 2.3 and 2.9 Bq m^{-3} from the southern to the northern part of the Aegean Sea. The respective values of the Cyprus Sea and the Levantine Basin were lower [228, 230, 233, 237].

Considering the generic inventory of ^{137}Cs in the Black Sea -Aegean Sea during 1993, the average concentration of ^{137}Cs in sea water in the Black Sea varied up to 130 Bq m^{-3}. The respective value in the mouth of the Dardanelles was 120 $\pm$ 8 Bq m^{-3}, whereas the estimated mean values for the Aegean Sea was 20.7 $\pm$ 14.7 Bq m^{-3} with the maximum value 9.3 $\pm$ 4.2 Bq m^{-3} [283]. This value was quite high, compared to the pre-accident levels 2.6 $\pm$ 0.3 Bq m^{-3} or to the respective value for the Ionian Sea 9.2 $\pm$ 2.5 Bq m^{-3}. As regards the concentrations of ^{137}Cs in the Cyprus Sea, the concentrations of ^{137}Cs were measured during the end of the warm period of 1998 in 15 stations in the Northeastern Cyprus Sea, under the frame of the NPPT-98 International programme. According to these measurements, the ^{137}Cs concentrations range to 0.3–3.0 Bq m^{-3} (with two values of about 5 Bq m^{-3}, excluded from the range, as they concern coastal stations influenced by land and strong evaporation) which are the background values measured before the Chernobyl accident [255, 266, 282].

Considering the radioactive pollution of the Black Sea from the Chernobyl fallout, the mouth of the Dardanelles has been determined as the impact source of the soluble artificial radio nuclides, which remain in the water column for a reasonable time and are transferred through the water mass exchange. Thus, we assume that a more or less predictable amount of ^{137}Cs is discharged to the Aegean Sea-Eastern Mediterranean basin during the purification processes of the Black Sea through the Dardanelles channel and dispersed to the South- Western Aegean Sea via surface current circulation, with respect to the sinking; procedure of the radioisotope.

As regards the pollution sources in the Eastern Mediterranean, Cyprus Sea—Southeastern Mediterranean is the less influenced area, as ^{137}Cs transferred from the Black Sea is widely dispersed and sinks in the Aegean Sea.

The natural occurring radioisotopes of uranium, radium, thorium and potassium are known to be present in the earth–crust and environment ever since its creation. The exposure to gamma radiation, which is emitted from these radioisotopes, could be significant for the public health, since the population is continuously exposed to such radiation. Even with the availability of many related studies, the general public is still not fully aware of the natural terrestrial origin radiation.

In Greece, the study on natural radioisotopes is limited. However, studies on coastal areas of volcanic island [233] and on the natural radiation [234] and elsewhere [235–237] have shown these radioisotopes to be present in sand samples. The beaches of Pieria Perfecture, in Greece, have always been popular for out-door activities during the summer time. This coastline is of high touristic interest, with four main recreational spots: Korinos, Katerini, Litohoro and Leptokaria. This paper reports on the natural radionuclide content in the beach sand collected from 9 sampling sites of the coastline of Pieria, starting at Korinos and ends at Leptokaria. The samples have been analyzed by using a high purity germanium gamma ray spectrometer, into low background chamber [282].

The sand samples were collected from nine selected locations along the coastline of Pieria Prefecture. At each location, 10 samples were taken along the mid way point between the points of high tide and low tide within meter interval. The samples were fug from 5 cm below the surface. They were then placed inside polythene begs with the content of each bag has a net weight of approximately 1 kg. These samples were then taken to the laboratory and dried into the oven at 110°C. They were than meshed to about 500 micron and stored in Marinelli beakers for about 3 weeks before measurement were taken using the HPGe gamma ray spectrometer [229].

The gamma spectroscopy system is consisting of a high purity coaxial Germanium detector p-type (CP 2100, Tennelec). The sample chamber was a cylinder 12 cm in diameter and 25 cm in height and it was shielded by 5 cm of lead and 0.5 cm to copper in order to low the background. The full width at half

Table 3.54 Locations and hourly dose rate (nGy h^{-1})

Site	Location	Hourly dose (nGy h^{-1})
P1	Korinos 1	65
P2	Korinos 2	65
P3	Katerini 1	62
P4	Katerini 2	81
P5	Katerini 3	79
P6	Litohoro 1	76
P7	Litohoro 2	54
P8	Litohoro 3	51
P9	Leptokaria 1	50

maxima (FWHM) of the system was found to be 1.95 at 1332 keV of ^{60}Co. The linearity of the detector was checked with ^{152}Eu source and a simple regression analysis gave a straight line with a correlation coefficient of 0.999. The radionuclides used were supplied by "The Nucleus", Oak Ridge, USA. The counting time was 86,400 s for all the samples in order the time to be sufficient to measure almost any level of concentration [255, 266].

Table 3.54 shows the hourly radiation doses for the studied sites. The doses were found to range between 50 and 81 nGy h^{-1}.

The dose rate due to the external gamma radioactivity of terrestrial origin was evaluated by use of the following equation:

$$D = 0.010A_1 + 0.506A_2 + 0.284A_3 + 0.421A_4 + 0.045A_5$$

where D is the absorbed dose rate in nGy h^{-1} and A_1, A_2, A_3, A_4, A_5 are the specific activities (in Bq kg^{-1}) of the isotopes ^{238}U, ^{226}Ra, ^{228}Ra, ^{228}Th and ^{40}K respectively [238].

The annual dose equivalent (H) in μSv y^{-1} is given by:

$$H\ (\mu Sv\, y^{-1}) = 6.13D\ (nGy\, h^{-1})$$

Table 3.55 shows the average activity concentration of the beach sand sampling from the studied nine locations, expressed in Bq kg^{-1}. Generally, the magnitude of the concentration is in the order of K > Th > U>Ra [253, 254]. For uranium and thorium activities, concentration readings obtained from different gamma energies show slight variation. However, both readings are of the same order and within the limits of the experimental error. The only exception is for the Thorium concentrations for the locations of Korinos and Katerini (P1, P3 and P4). Areas around Korinos (locations P1 and P2) and Katerini (P4 and P5) consistently gave rather high concentrations of Thorium. Concentrations of Thorium between 1 and 2 ppm and uranium between 0.3 and 0.6 ppm suggested that beach sand might contain minerals such as monazite [239].

Table 3.55 Average concentrations of U, Ra, Th and K

Geographic region	^{238}U (Bq kg^{-1})	^{226}Ra (Bq kg^{-1})	^{226}Ra (Bq kg^{-1})	^{228}Th (Bq kg^{-1})	^{40}K (Bq kg^{-1})	Annual dose equivalent (μSv y^{-1})
P1	47 ± 6	20 ± 3	32 ± 7	46 ± 5	581 ± 31	400 ± 50
P2	41 ± 5	28 ± 2	36 ± 9	48 ± 7	437 ± 29	396 ± 45
P3	63 ± 7	30 ± 3	44 ± 10	32 ± 9	441 ± 18	378 ± 35
P4	44 ± 4	38 ± 4	67 ± 12	48 ± 9	494 ± 33	497 ± 70
P5	38 ± 3	47 ± 6	36 ± 9	44 ± 7	588 ± 57	487 ± 80
P6	33 ± 3	64 ± 5	25 ± 14	38 ± 9	446 ± 63	465 ± 70
P7	28 ± 2	24 ± 7	39 ± 9	26 ± 7	432 ± 68	330 ± 40
P8	26 ± 3	28 ± 3	34 ± 7	25 ± 9	362 ± 70	312 ± 35
P9	26 ± 4	24 ± 7	28 ± 6	29 ± 6	377 ± 50	304 ± 40

Each value represents the average value of the measurements of ten samples per location. Locations around Litohoro and Leptokaria (P7, P8 and P9) showed low concentration of uranium, thorium and potassium.

Conclusions

The present studies shown that the natural radioactivity derived from uranium, thorium and potassium are below the maximum permissible dose of 1.5 μSv per year, which is equivalent to 370 Bq kg^{-1}. The studied isotope concentrations are in good agreement with the population–weighted mean value of Greece (380 μSv y^{-1}) [238] and with elsewhere concentrations [240, 257]. Nevertheless, in the international literature maxima exceeding 5,000 μSv y^{-1} have been reported for other regions of the world [241]. It is concluded that the beaches along Pieria Prefecture of Greece are low in natural radioactivity and quite safe for any recreational activity [263].

3.3.8 *Comparative Studies Concerning ^{137}Cs Effect on the Biological Status of Carp* (**Cyprinus carpio**) *and Eel* (**Anguilla anguilla**)

The results obtained from the studies carried out are comparative for the accumulation of ^{137}Cs in two fish species different in their anatomical and physiological functional behavior [237, 280, 281]. The purpose is to find their sensitivity to ^{137}Cs in order to take steps in case of danger of radioactive contamination of waters. There are no found visible or microscopic changes in the behavior of the control fishes. The respiration balance of both fish species is normal. The results show that the amount of ^{137}Cs in the muscular tissue of the carp is higher compared to that of the eel.

Table 3.56 ^{137}Cs concentration (Bq/kg) in the muscular tissue of cultured fresh-water fish: *Cyprinus carpio* and *Anguilla anguilla*

Time of killing (days)	*Cyprinus carpio*		*Anguilla anguilla* (3000 Bq/l)[a]
	(1500 Bq/l)[a]	(3000 Bq/l)[a]	
15	980	1150	1170
30	2980	6437	1000
45	3550	8023	2500
60	–	10460	4000
90	4985	12141	4400
120	10850	18959	4900

[a] 1500 Bq/l and 3000 Bq/l are the ^{137} Cs concentrations in the tank water

As Table 3.56 indicates, the accumulation of ^{137}Cs in the eel is progressive for a period of two months. It seems that after the above-mentioned period the amount of cesium is stabilized. The slime, which is destroyed in the presence of salt or is removed, in one percent solution of ammonia. It may play a protective role against the absorption of ^{137}Cs from the skin of eels. The secretions of mucous membranes have important protective and lubricate functions. Their active substances are glycoproteins that contain numerous negatively charged oligosaccharide chains.

Eels are smoothing skinned and very slimy. Most people would say that an eel has no scales, but actually many very small scales are present but they are embedded under the skin. Due to the absence of large scales, an eel can breathe through its skin as well as through the gills. The proportion of breathing carried out through the gills is about 40 percent and that through the skin is about 60 percent. This means that the eel in comparison to carp takes up less water. That's why higher amounts of ^{137}Cs were found in carp and the accumulation was related to the time of exposure. In addition, the polyelectrolytic behavior of slime together with the rate of respiration and the different compositional characteristics of the fish organs can explain the differences in accumulation and dissimilar distribution of ^{137}Cs in both species. The histological studies revealed that the eel is more resistant to ^{137}Cs exposure than carp. The presence of ^{137}Cs in all organs tested caused a gradual degeneration. With high doses up to 300 Bq/l degeneration of epithelial cells and secondary fusion of lamellae final in the gills, are very common [231, 232].

It is concluded that allergic or toxic effects of ^{137}Cs caused liver hyperemia, hemorrhages in musculature fibers and focal degeneration of epithelial cells of the renal tubules.

Relevant photos are listed below (Figs. 3.30, 3.31, 3.32, 3.33).

Tables 3.57 and 3.58 present the results from the previous studies on concentration factors (CF) for the different classes of fresh-water organisms.

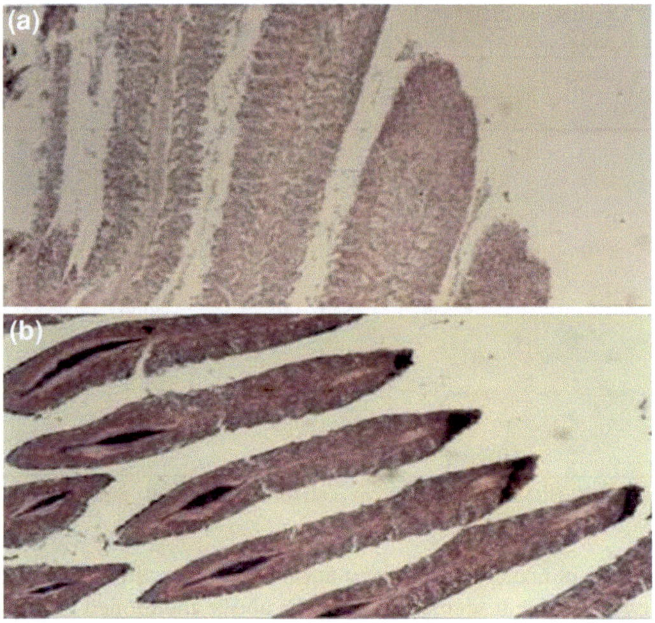

Fig. 3.30 Remarkable epithelial hyperplasis and fusion of some secondary lamellae in the gills. Final stage. **a** Carp (×250), **b** Eel (×100)

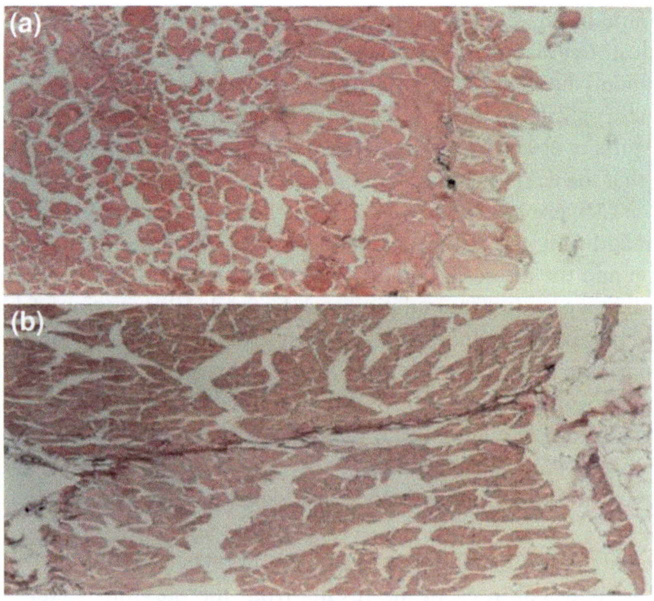

Fig. 3.31 Degeneration of muscles fibers. Final stage. **a** Carp (×250), **b** Eel (×100)

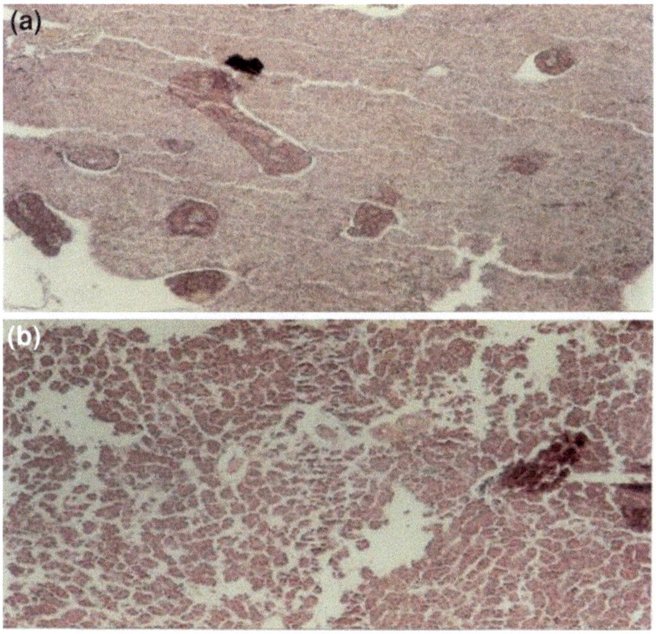

Fig. 3.32 Degeneration of hepatic cells. Final stage. **a** Carp (×250), **b** Eel (×100)

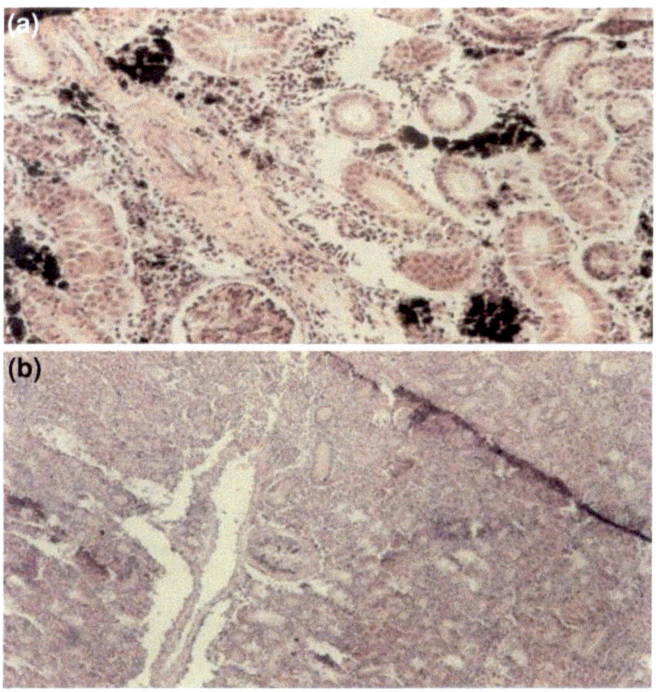

Fig. 3.33 Degeneration of Kidney parenchymal cells and hydeopsy of renal tubules. Final stage. **a** Carp (×250), **b** Eel (×100)

Table 3.57 Radionuclide dependent environmental parameters [287]

Nuclide	Sediment K_d (in m^3t^{-1})			Concentration Factor (in m^3t^{-1})		
	Coastal	Ocean	River	Fish	Crustaceans	Molluscs
^{3}H	1	1	0	1	1	1
^{54}Mn	2.00E + 05	2.00E + 08	1.00E + 04	400	500	5000
^{58}Co	2.00E + 05	1.00E + 07	1.00E + 03	1000	5000	5000
^{60}CO	2.00E + 05	1.00E + 07	1.00E + 03	1000	5000	5000
^{65}Zn	2.00E + 04	2.00E + 05	1.00E + 03	1000	50000	30000
^{90}Sr	1.00E + 03	2.00E + 02	2.00E + 03	2	2	1
^{95}Zr	1.00E + 06	5.00E + 05	6.00E + 04	20	200	5000
^{99}Tc	1.00E + 02	1.00E + 02	2.00E + 02	30	1000	1000
^{103}Ru	3.00E + 02	1.00E + 03	1.00E + 03	2	100	2000
^{106}Ru	3.00E + 02	1.00E + 03	1.00E + 03	2	100	2000
^{110}Ag	1.00E + 03	1.00E + 04	2.00E + 02	500	5000	10000
^{124}Sb	1.00E + 03	5.00E + 02	1.00E + 03	400	400	200
^{125}Sb	1.00E + 03	5.00E + 02	1.00E + 03	400	400	200
^{133}Ba	5.00E + 03	1.00E + 04	0	10	1	20
^{134}CS	3.00E + 03	2.00E + 03	3.00E + 04	100	30	30
^{137}CS	3.00E + 03	2.00E + 03	3.00E + 04	100	30	30
^{144}Ce	2.00E + 06	1.00E + 08	1.00E + 04	50	1000	5000
^{239}U	1.00E + 03	5.00E + 02	1.00E + 03	1	10	30
^{239}Pu	1.00E + 05	1.00E + 05	1.00E + 05	40	300	3000
^{239}Pu	1.00E + 05	1.00E + 05	1.00E + 05	40	300	3000
^{241}Am	2.00E + 06	2.00E + 06	4.00E + 05	50	500	20000
^{244}Cm	2.00E + 06	2.00E + 06	4.00E + 05	50	500	30000

Table 3.58 Activity concentrations of natural radionuclides in Aegean Sea (mBq l^{-1}) [288, 289]

	^{238}U	^{226}Ra	^{232}Th	^{40}K	^{228}Ra	^{228}Th
Station 1[a]						
MV±SD	38 ± 4	1.52 ± 0.07	0.024 ± 0.007	10641 ± 793	2.91 ± 0.6	0.10 ± 0.01
Min	32	1.46	0.018	9811	8	0.08
Max	42	1.63	0.028	11834	2.08	0.11
					3.92	
Station 2[a]						
MV±SD	39 ± 3	1.53 ± 0.11	0.020 ± 0.055	10651 ± 813	3.41 ± 0.5 0.	11 ± 0.02
Min	38	1.40	0.015	9875	5	0.07
Max	43	1.67	0.027	11847	2.96	0.13
					4.32	
Station 3[a]						
MV±SD	39 ± 3	1.50 ± 0.08	0.022 ± 0.005	10631 ± 741	3.22±0.6	0.090 ± 0.008
Min	35	1.39	0.017	9890	7	0.08
Max	43	1.59	0.029	11679	2.17	0.10
					4.01	
Station 4[a]						
MV±SD	37 ± 4	1.56 ± 0.04	0.022 ± 0.004	10689 ± 795	3.12±0.5	0.082 ± 0.019
Min	33	1.53	0.016	9895	1	0.06
Max	43	1.61	0.028	11789	2.67	0.11
					3.97	
Station 5[b]						
MV:l:SD		1200 ± 1000		19000 ± 600	600 ± 600	
Min		<100		19000	<100	
Max		1900		20000	1100	

[a] [288]
[b] [289]

Conclusions

The presence of ^{137}Cs causes allergic and toxic effects [267–269]. The ability of both species to accumulate great amounts of ^{137}Cs makes them very precious biological indicators of radioactivity. The results from these studies emphasize the necessity of study on more aquatic species and their sensitivity to ^{137}Cs, so that preventive steps can be taken in case of a nuclear accident or as a result of radionuclides release in the environment. More studies should be made on ^{137}Cs uptake considering different physical factors such as pH, ion exchange, differences in concentration, etc.

Chapter 4
General Discussion on the Results

The results obtained from the studies carried out present information of great interest about the radioactive contamination in Greece. In addition, correlations were found between radioactive elements and nuclides of great interest upon reprocessing the data. There has been demonstrated the dependence of their content and half-life on atmospheric physical parameters (temperature, air speed, relative humidity), composition and structure of the soil, reology of the basic substances of foods and plants, active acidity (pH), etc.

On the ground of the conclusions obtained concerning the behavior of radioactive isotopes under different conditions, lines for actualization were found of the technological processes upon soil cultivation, plant growing and production of the main foodstuffs.

The studies were made in the territory of Greece not only because of the nationality of the author. The "cloud" of radioactive substances as a result of the Chernobyl accident (1986) has mostly affected Greece, especially its northern area. The total fallout of radioactive dust that was measured in Thessaloniki was found to be 150 kBq/m^2.

Studies on the natural and anthropogenic radioactivity show that the atmospheric and soil radioactivity has been increased about 2.5 times, as a result of secondary radioactive exposure of the troposphere. Thus the yield that had been planted in 1986 has been contaminated mainly through its root system.

The most dangerous contaminants were found to be cesium (^{137}Cs and ^{134}Cs). That is why their content is the main subject of the research work in the dissertation paper. Due to the long half-life of ^{137}Cs (30 years), its amount remains relatively constant. It was found that cesium, opposite to potassium, enters the biological cycle of plants mainly through their leaves and less through their roots.

By calculating the concentration of ^{137}Cs and ^{40}K in Greek soils, data has been obtained that characterizes radioactive fallout in Greece, not only in 1986 but also a decade after it. The mean deposition of ^{137}Cs in 1986 was measured to be 0.01—13.7 kBq/m^2 and 10 years after the accident at Chernobyl it reached the value of

F. K. Vosniakos, *Radioactivity Transfer in Environment and Food*,
Environmental Science and Engineering, DOI: 10.1007/978-3-642-28741-1_4,

0.4–14.4 kBq/m^2. Regional studies have shown that Western Macedonia, Northern Thessalia and Central Macedonia are mostly contaminated. Sixty four percent of the Greek land has depositions between 0 and 3.5 kBq/m^2 and 11 % of Greek land has depositions higher than 12.0 kBq/m^2. Radioactive contamination in Greece is 2–3 times higher compared to that in England, Germany and Denmark.

^{40}K concentration for 1996 was found to be between 5.1 and 16.5 kBq/m^2. Higher values for ^{40}K have been found in the islands. This can be explained by the geological characteristics of their soils, taking in consideration that ^{40}K is a natural radionuclide. The values found were in compliance with other measurements [13].

The results obtained for ^{40}K are an indicator of the total concentration of potassium in the soil. It has been demonstrated that potassium has the same size and properties like cesium. Because of this, areas with higher concentration of ^{40}K are less affected by those with ^{137}Cs.

Plants absorb less cesium when there is enough potassium in their soil. The high concentration of cesium in Greek soils is explained mainly by their mineral composition. The soils in Northern Greece are with high presence of clay with vermolite (from the mica family), which is the dominant material. Clay is 49% of the soil composition, silt is 37% and the organic substances are 1.1%. In addition, Greek soils have a pH between 7 and 8 and high values of potassium (from 220 ppm). The results aforesaid are in compliance with other authors. These parameters of the Greek soils help retaining cesium and its slow penetration in soil depth, which for the last period was found to be from 0.5 to 5.0 cm.

It was found that ^{134}Cs and ^{40}K have one and the same oxidation number and equal diameter of particles. The absorption of cesium from a given plant is lower than the absorption of potassium, probably due to the lower contiguity of cesium with the carriers of cell membranes of the plants roots. Another reason is that the concentrations of ^{40}K are one order higher than those of ^{137}Cs. In addition, cesium must connect with the stable potassium, which has higher concentrations. The total potassium presence is 1% W/W. The fixing factors of cesium retaining in the soil do not favor its absorption from the plants.

The transport of different radionuclides from soil to plant depends on the soil properties, the type of plant and the vegetation variety, the climatic conditions and the way of the soil cultivation. Some of the most important soil properties are its structure, pH, the exchangeable potassium and calcium, the type and amount of clay and the presence of organic substances.

According to the results obtained, the factors that affect ^{137}Cs transfer from soil to plants (CR_1) vary between 0.0486 and 0.1092 (0.0789 on the average). While the factors that affect ^{40}K transfer (CR_2) vary between 0.5859 and 0.6662 (0.6215 on the average). It was found that the most frequently measured concentration of ^{137}Cs in the plants was 2.47 kBq/kg dry plant, and in clover was 5.54 kBq/kg dry plant. The concentration values found for ^{40}K radioactivity are 0.062 for the plants and 0.989 for clovers. This confirms the high presence of ^{40}K in soil as well as in plants.

Upon soil analysis, there has been applied principally systematical method for estimating the concentration level in a given territory. After that the ways of radionuclide deposition in the soil have been determined and the radioactivity of the ground surface and soil has been measured.

Upon study of the radioactivity transfer in soil, objects of different utilization, which were located at a certain distance from one another, have been chosen (Agios, Airport, and Drama). On the basis of the results obtained from their study the transfer factors were clearly formulated: potassium presence in the soil; amount of clay and mica in the soil; amount of organic matter, pH.

The correlation found between ^{137}Cs transfer and the natural calcium and magnesium in the soils, proved to be the greatest achievement of this dissertation.

The transfer of ^{137}Cs and ^{40}K from soil to plants was studied on crops, which had been most widely used as food for animals—corn, reed, and rice. The reciprocal correlation in the transfer of ^{137}Cs and ^{40}K were determined to be linear with coefficient 0.66. In a similar way the correlation was presented between the transfer factor of ^{137}Cs and ^{40}K with regard to their presence in the soil by using the corn as a sample. It has been demonstrated that the uptake of ^{137}Cs from corn in 1988 has not changed compared to that in 1986. It seems that the fixing of ^{137}Cs in the soil does not affect the exposure to the extent it was considered 10 years ago. It is evident that there is a necessity for the analyses to be repeated in order to create a stable model of determining the radioactivity movement. Based on the results obtained from the study on atmospheric radioactivity and the radioactivity that had been transferred from the atmosphere to the soil and plants, it becomes obvious that finally most radionuclides transfer to man's food.

The research work concerning milk, milk products and freshwater fish is advisable. Contamination with ^{137}Cs and ^{131}I is most expectable in the foods aforesaid. It has been demonstrated and scientifically based that upon processing milk into cheese and curd, there possibilities appear for removal of a great amount (70 ± 10%) of the radioactive elements through the whey. The attempts at determining the conditions affecting the rate of radioisotopes transport—their nature, the composition of the starting material (milk or whey), etc., are the great contributions of this project. On the basis of this, there are given instructions for safe production of the studied products, for example by changing the heating-processing regime, the extent of serum removal, etc.

The results concerning the impact of radioactive contamination of acid-lactic products with ^{131}I lead not only to application of methods for decreasing the radioactive danger in these products. Possibilities appeared of stimulating the development and viability of the typical micro-flora of yogurt. On the other hand, a model has been developed of purifying the milk from radioactive elements, the application of which is very practical.

It has been demonstrated that together with the character of contamination, the fish species studied (*Cyprinus carpio* and *Anguilla anguilla*) may serve as important biological indicators of radioactivity. Studies on other species are required.

The approach of interpreting the results is not only practical. It has mainly a creative character, due to which new necessities and possibilities for future research work on the chosen line of the dissertation paper were found.

4.1 The Fukushima Case

Nowadays, the Fukushima accident in Japan (2011), having the same size of magnitude in the INES scale as the Chernobyl accident, can be considered to be quite similar: the radiation levels were in the ranged 35–150 mSv/h in some spots and overall at 1–10 mSv/h. Therefore, all the consequences from the Chernobyl accident described in detail before are similar in Fukushima as well. Especially the contamination of the soil through the air pollution deposit can affect directly and indirectly the food chain. The radioactivity transfer in the food chain but also in the soil is a threat for the coming years. Additionally to the soil contamination also the marine environment was affected. On the other hand, there is a huge quantity (about 90,000 tones) of contaminated water coming from the reactors which also has to be treated in a special way. Therefore, for each reactor of Fukushima it is necessary:

(1) In the reactor Number 1 the fuel rods completely melted after burning through the pressure vessel surrounding them, and a part leaked. Radioactive water has to be treated as soon as possible.
(2) The condition of the crane related to the fuel in reactor Number 2 have to be examined. Here the fuel rods partially melted. But there are radioactive water leaks too.
(3) In reactor 3, the cover has to be restored and remove contaminated waters. Here the fuel rods partially melted.
(4) For reactor 4, a crane should be build, and the remaining fuel be collected.
(5) For reactors 5 and 6, there is no need for any intervention because they were successfully shut down before the accident.

Chapter 5
Basic Conclusions

The results obtained from the research carried out about the radioactive transfer air, soil, plants, and food give ground to the following basic conclusions:

1. In the atmosphere in Greece there have been identified the radionuclides Pb, Ba, Sr, Kr, Ru, K, Cs, the concentration of which vary from 25 to 1,450 Bq/m^3
2. It has been found that the radioactive contamination in the region during the period 1986–1999 has increased by 2.1–1.5 times and the diffuse radioactivity of areas with comparatively low radiation has increased after the accident at the nuclear Power Station in Chernobyl by 400–800%.
3. A linear correlation has been demonstrated between the radioactive nuclides concentration and the physical parameters of the air (relative humidity, speed and temperature).
4. The transfer of the nuclides ^{137}Cs and ^{40}K from the air to soils and plants has been found.

    4.1. Alkaline soils (pH 7.8) retard ^{137}Cs transfer.
    4.2. ^{137}Cs uptake from the air is inversely proportional to the presence of ^{40}K.
    4.3. The transfer of ^{137}Cs and ^{40}K in plants depends more on the presence of potassium and calcium and less on the presence of magnesium.
    4.4. Plants with high presence of ^{40}K are poor in ^{137}Cs.
    4.5. The main Greek building material meet the limit of radiation dose to humans (1.5 mSv/year).

5. There has been found the mechanism of transport of the isotopes ^{131}I and ^{137}Cs from cow milk-to-milk products.

    5.1. The radioactive isotopes ^{131}I and ^{137}Cs pass from the contaminated milk in the products prepared from it through the water phase. The radionuclide ^{137}Cs has a greater transportability (82.4%) compared to ^{131}I (71.1%) under equal conditions.

F. K. Vosniakos, *Radioactivity Transfer in Environment and Food*,
Environmental Science and Engineering, DOI: 10.1007/978-3-642-28741-1_5,

5.2. The diffusion coefficient of radionuclides in cheese immersion in brine of NaCl and KCl or in water, depends on the solution concentration, the time of salting and the cheese structure.

6. It has been demonstrated that the radionuclide ^{131}I decreases the viability of *L. bulgaricus* and *S. thermophilus* in the yogurt by 45%.
7. The capacity of fresh-water fish has been found: *Cyprinus carpio* and *Anguilla anguilla* to accumulate large amounts of ^{137}Cs in their muscular tissue—10,850 Bq/kg and 4,500 Bq/kg respectively.
8. Models and mathematical equations, characterizing radioactivity and its transfer have been developed.

   8.1 A model has been developed of horizontal and vertical Gaussian distribution of airborne radioactive substances.
   8.2 A model has been developed of radioactivity migration in air, soil, and food.
   8.3 A model has been developed of purification of milk from ^{131}I and ^{134,137}Cs by ion-exchanging resins, applicable in practice.
   8.4 It has been shown clearly that a dose of 640 Krad at a temperature of 11 °C matures the cheese earlier (from 15 to 11 days).

9. Recommendations concerning practice:

The technological approaches upon cultivation of soil, cultivation of plants and reprocessing of the raw materials into foodstuffs shall be in compliance with the radioactive contamination of the environment.
Higher temperature regimes shall be applied upon pasteurization of milk.
When preparing cheese from milks, contaminated with radionuclides, larger amount of whey shall be removed from the cheese upon pressing and curd shall have lower water content.
The concentration of NaCl and KCl in brine shall be increased in case of increased radioactive contamination of milk.
Radioactive elements in the soils and especially the transfer factors of ^{137}Cs and ^{40}K shall be regularly tested.
Fresh water fish *Cyprinus carpio* and *Anguilla anguilla* shall be used as biological indicators of radioactive status of the fish and waters.
The presence of Cs in corn and other crops used for animal feeding shall be studied extensively.

Appendix A
Radioactive Concentrations in Geographic Periferials of Greece

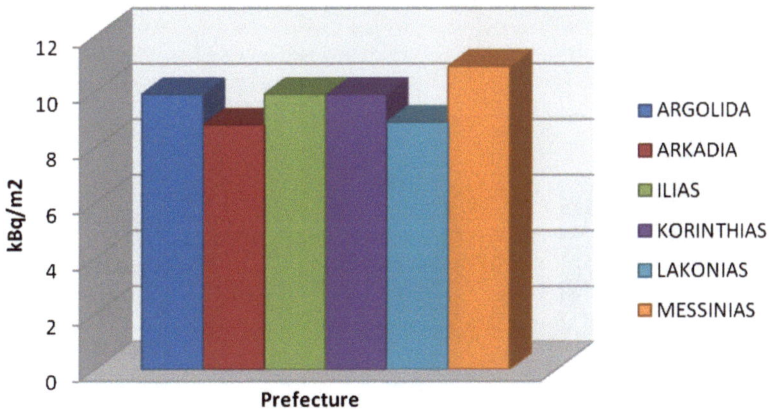

Fig. A.1 Radioactive concentration of K-40 in the periferial of Peloponisos

F. K. Vosniakos, *Radioactivity Transfer in Environment and Food*,
Environmental Science and Engineering, DOI: 10.1007/978-3-642-28741-1,

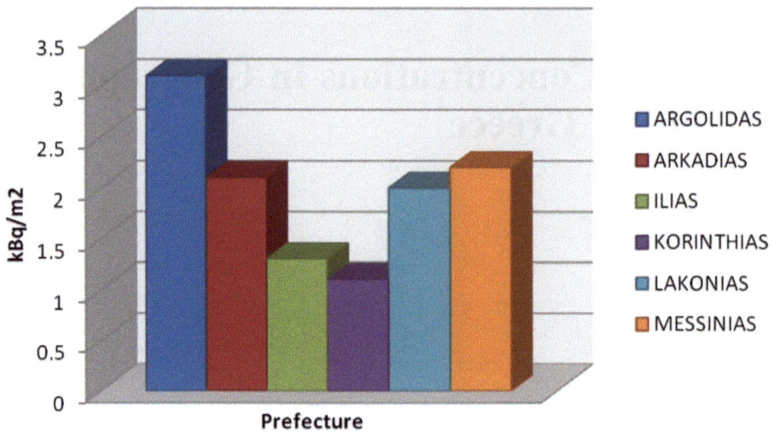

Fig. A.2 Radioactive concentration of Cs-137 in the periferial of Peloponisos

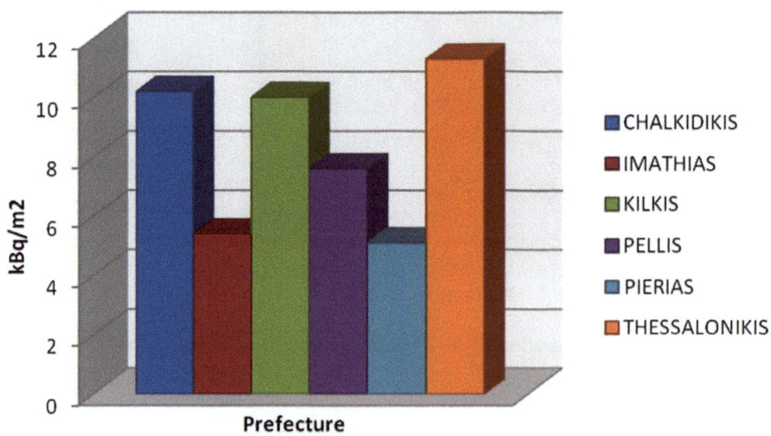

Fig. A.3 Radioactive concentration of K-40 in the periferial of central Macedonia

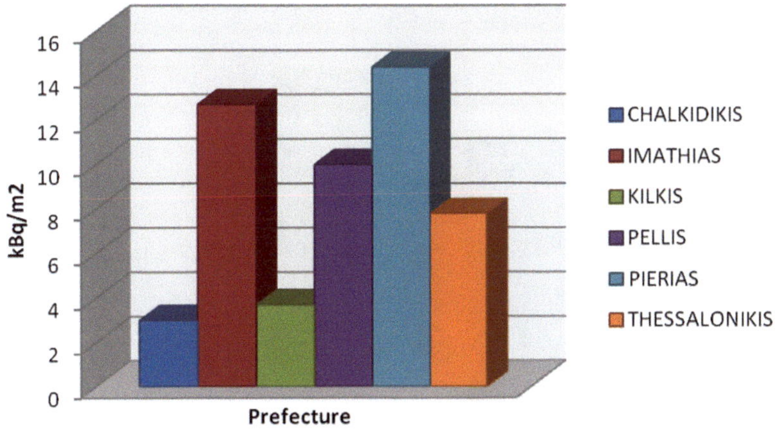

Fig. A.4 Radioactive concentration of Cs-137 in the periferial of central Macedonia

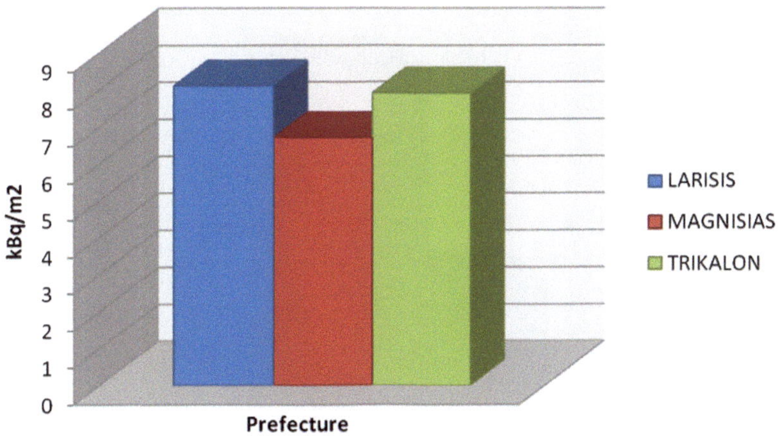

Fig. A.5 Radioactive concentration of K-40 in the periferial of Thessalia

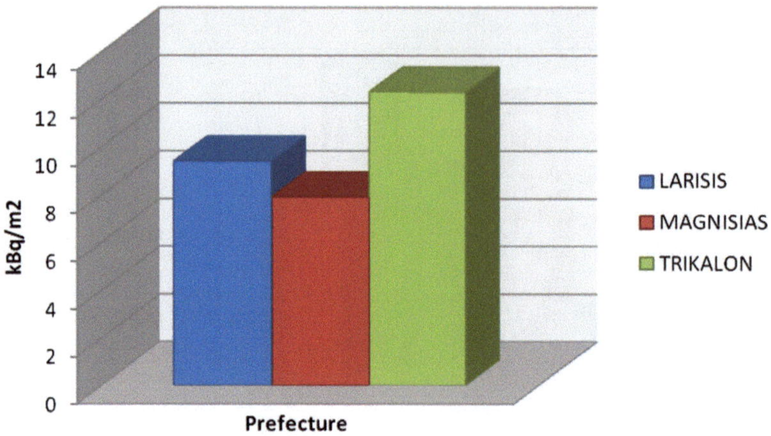

Fig. A.6 Radioactive concentration of Cs-137 in the periferial of Thessalia

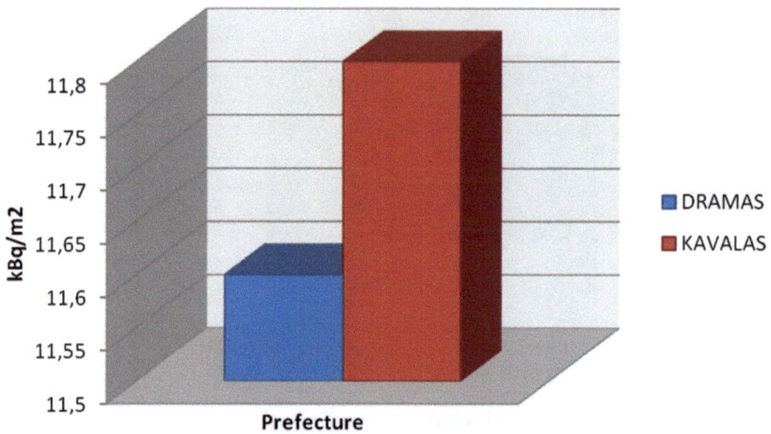

Fig. A.7 Radioactive concentration of K-40 in the periferial of East Macedonia

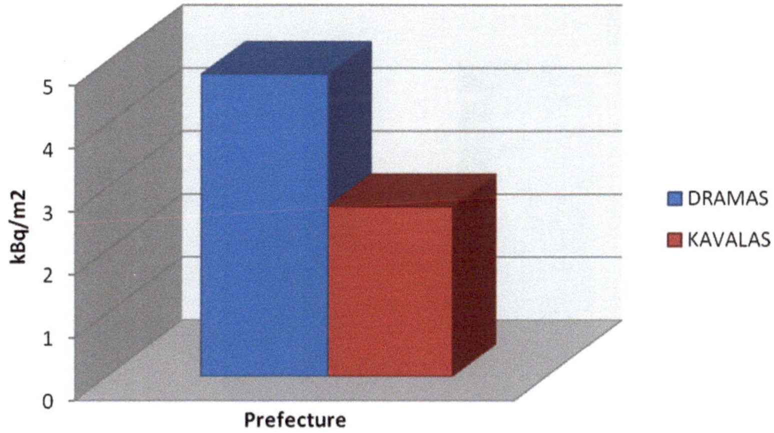

Fig. A.8 Radioactive concentration of Cs-137 in the periferial of East Macedonia

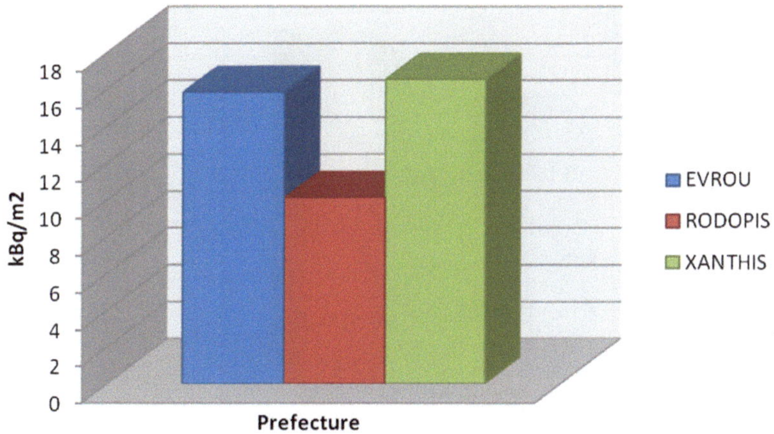

Fig. A.9 Radioactive concentration of K-40 in the periferial of Thrace

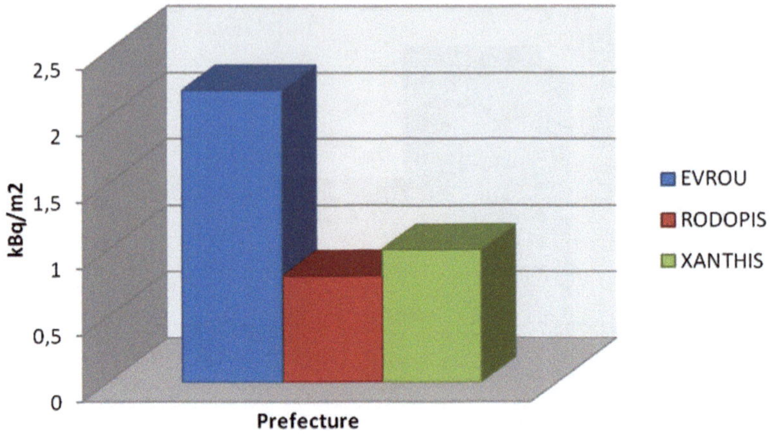

Fig. A.10 Radioactive concentration in the periferial of Thrace

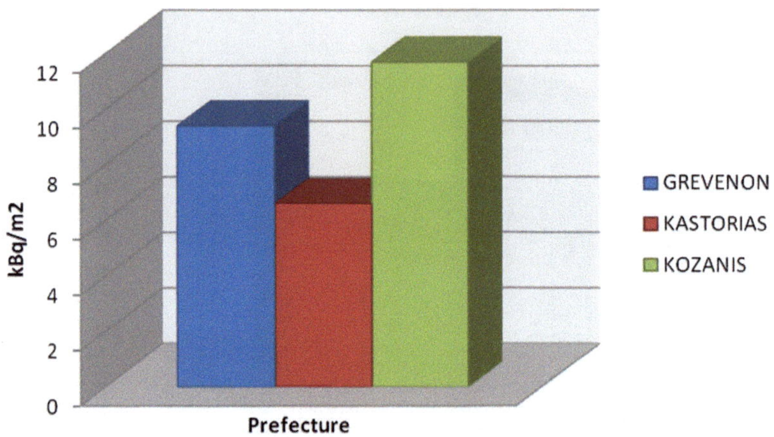

Fig. A.11 Radioactive concentration of K-40 in the periferial of West Macedonia

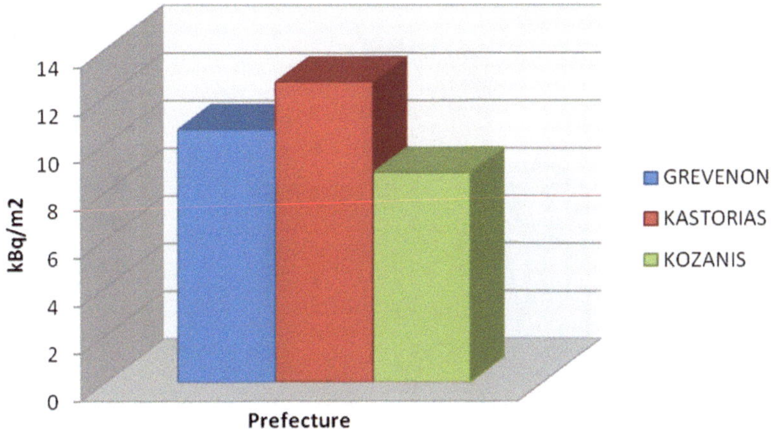

Fig. A.12 Radioactive concentration of C-137 in the periferial of West Macedonia

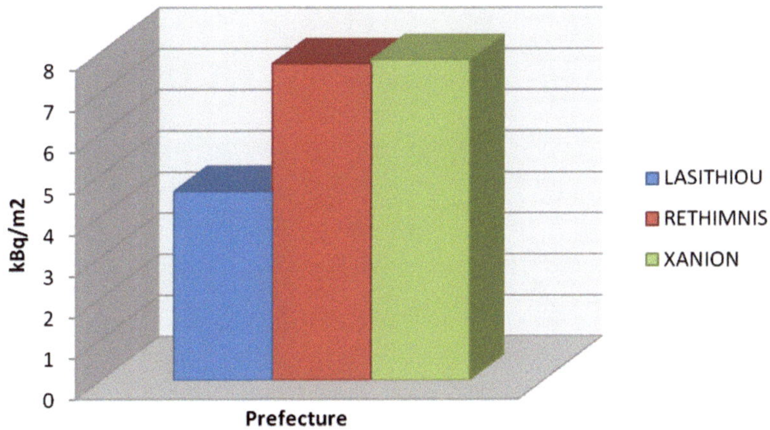

Fig. A.13 Radioactive concentration of K-40 in the periferial of Crete

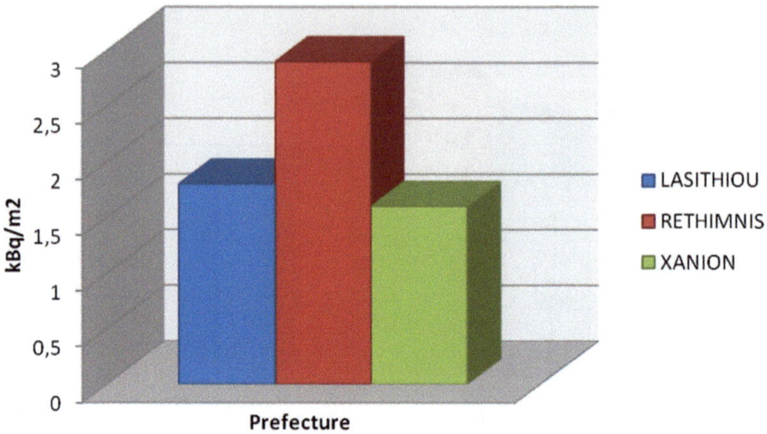

Fig. A.14 Radioactive concentration of C-137 in the periferial of Crete

Appendix B
Radioactive Concentrations in the Prefectures of Greece

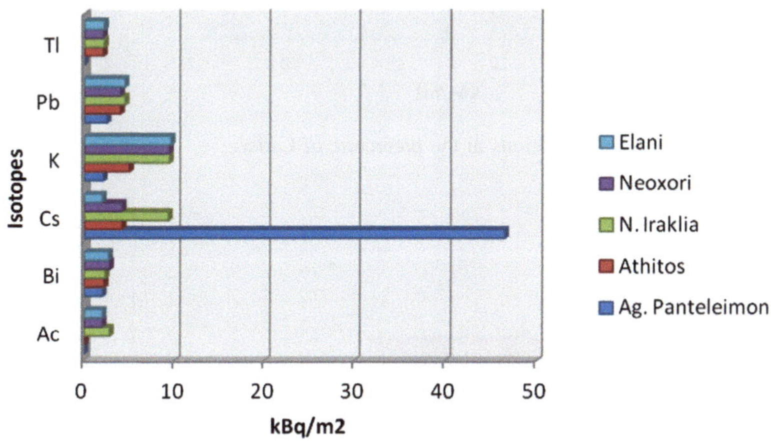

Fig. B.1 Radioactive concentrations in the prefecture of Chalkidikis

F. K. Vosniakos, *Radioactivity Transfer in Environment and Food*,
Environmental Science and Engineering, DOI: 10.1007/978-3-642-28741-1,

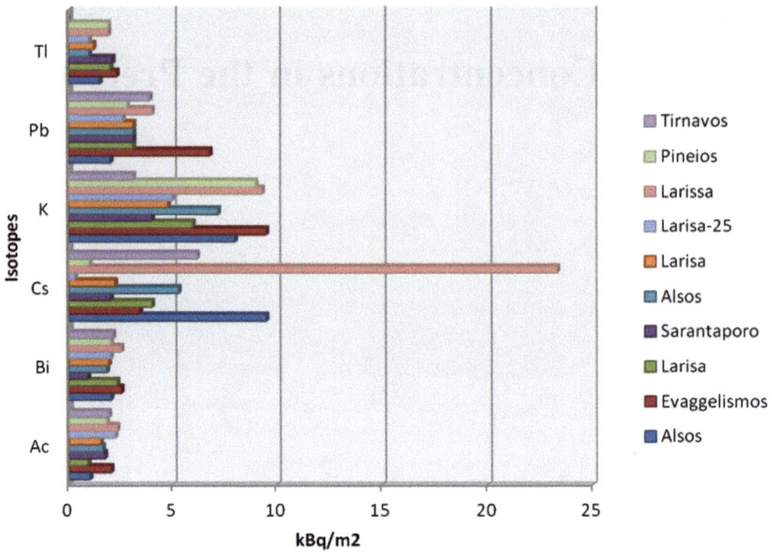

Fig. B.2 Radioactive concentrations in the prefecture of Larisis

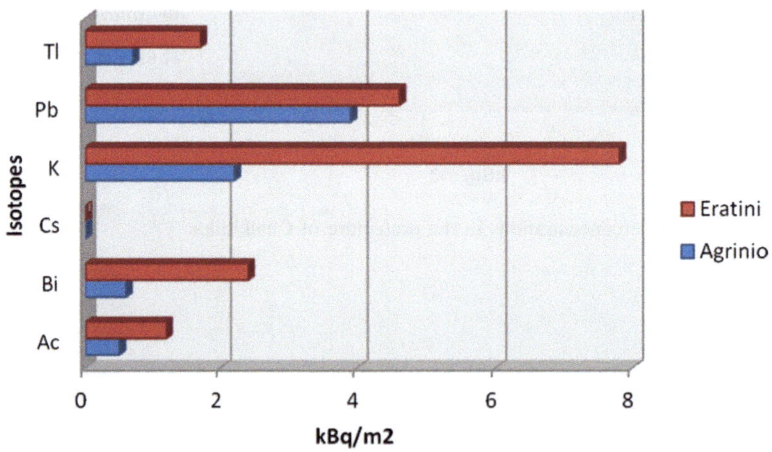

Fig. B.3 Radioactive concentrations in the prefecture of Etolo/nias

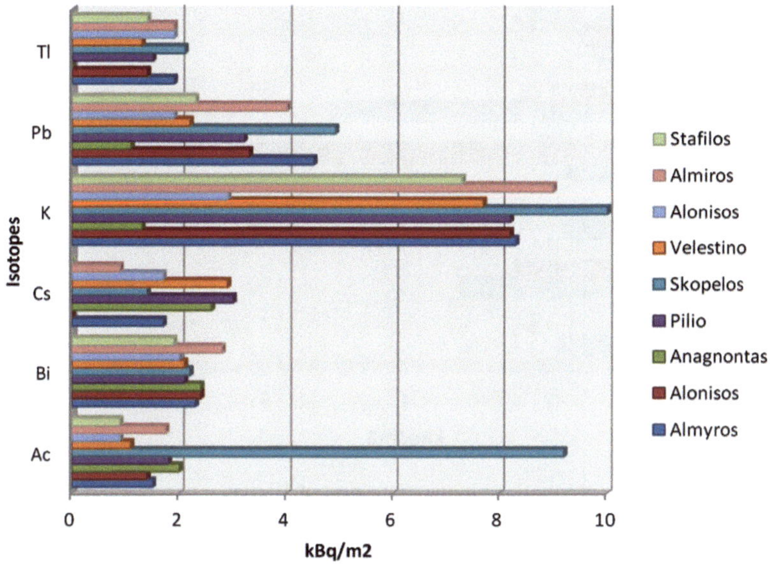

Fig. B.4 Radioactive concentrations in the prefecture of Magnisias

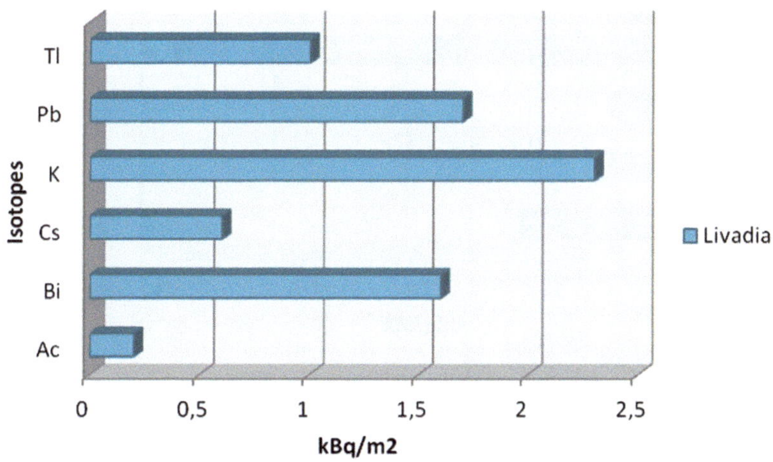

Fig. B.5 Radioactive concentrations in the prefecture of Fthiotidas

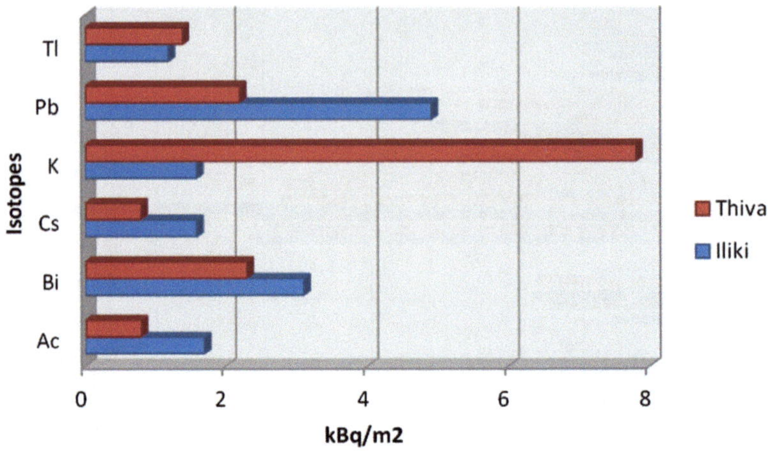

Fig. B.6 Radioactive concentrations in the prefecture of Viotias

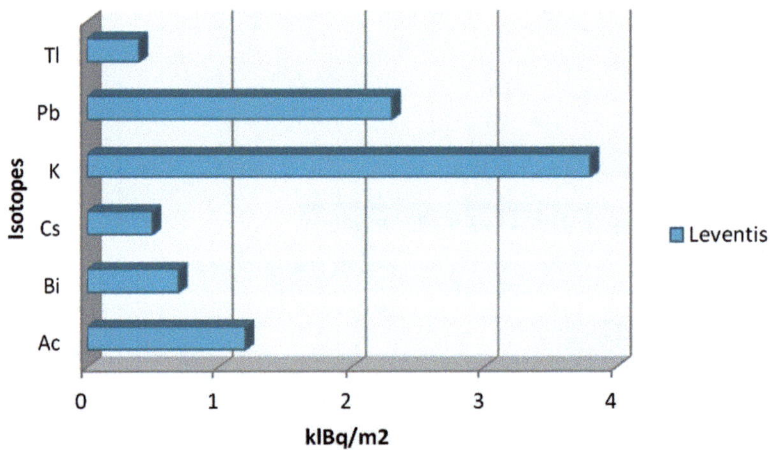

Fig. B.7 Radioactive concentrations in the prefecture of Attikis

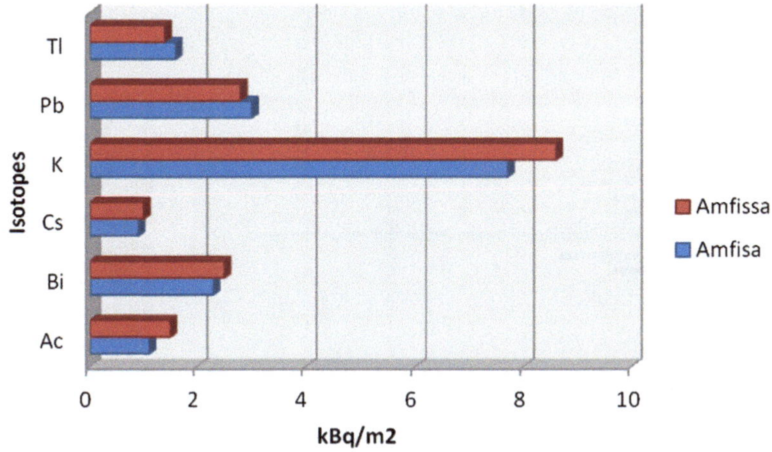

Fig. B.8 Radioactive concentrations in the prefecture of Fokidos

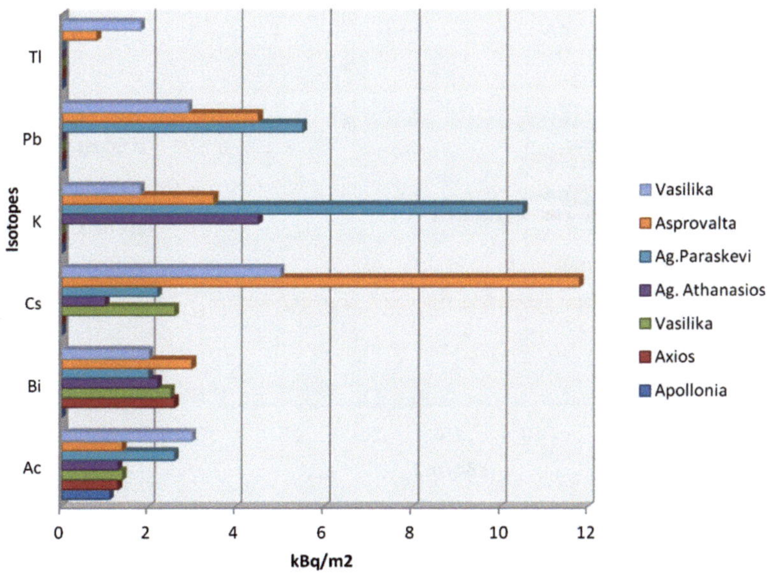

Fig. B.9 Radioactive concentrations in the prefecture of Thessalonikis

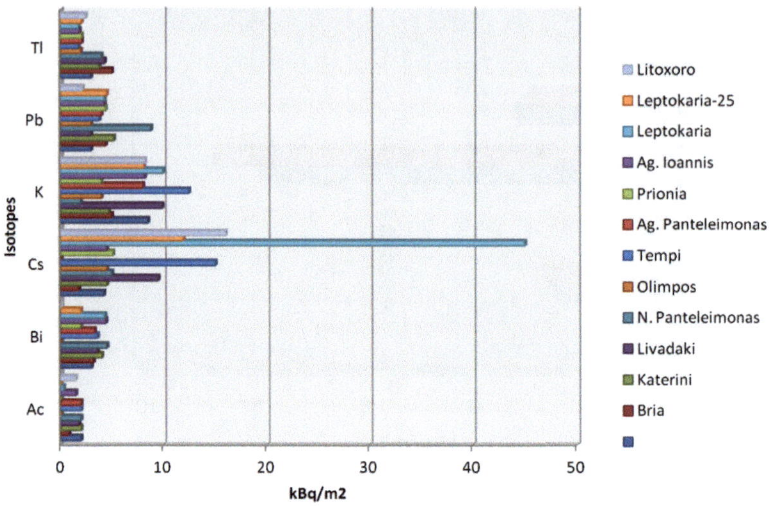

Fig. B.10 Radioactive concentrations in the prefecture of Pierias

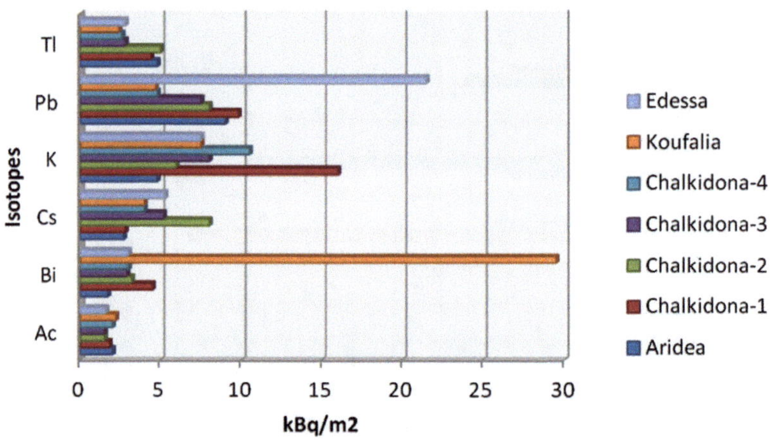

Fig. B.11 Radioactive concentrations in the prefecture of Pellas

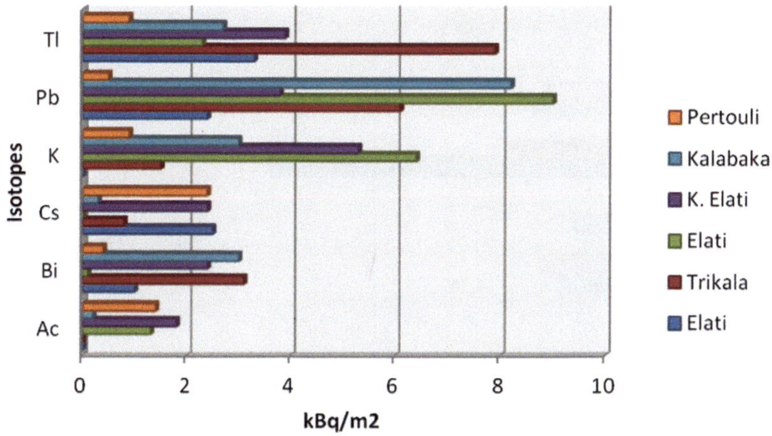

Fig. B.12 Radioactive concentrations in the prefecture of Trikalon

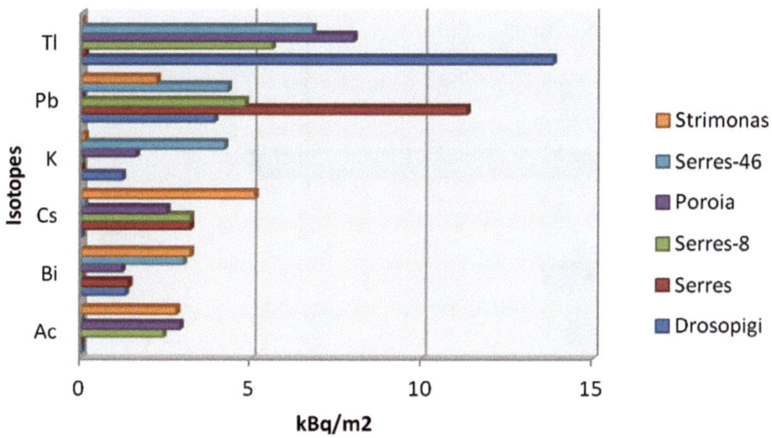

Fig. B.13 Radioactive concentrations in the prefecture of Seron

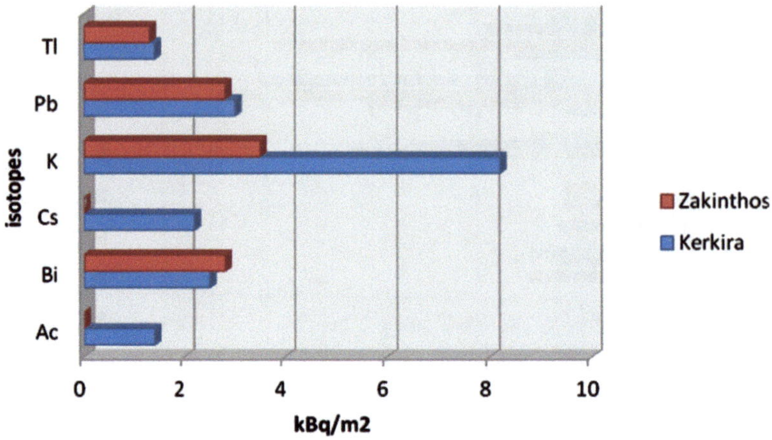

Fig. B.14 Radioactive concentrations in the prefecture of Ionion

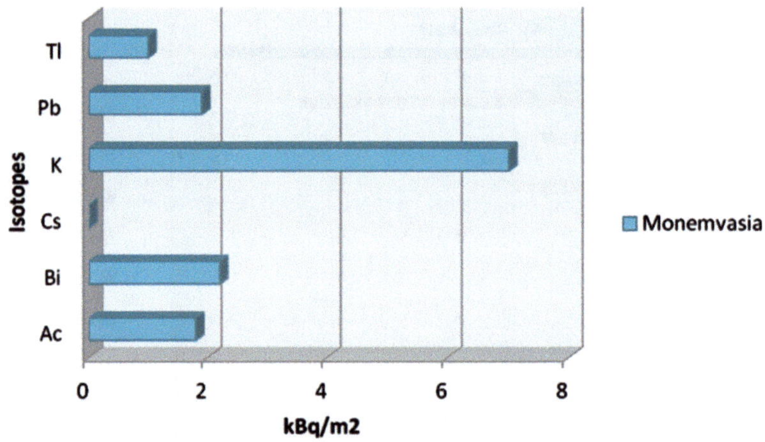

Fig. B.15 Radioactive concentrations in the prefecture of Lakonias

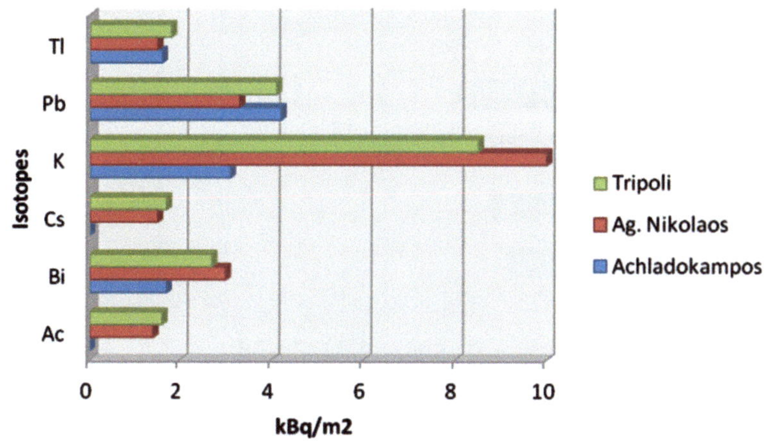

Fig. B.16 Radioactive concentrations in the prefecture of Arkadias

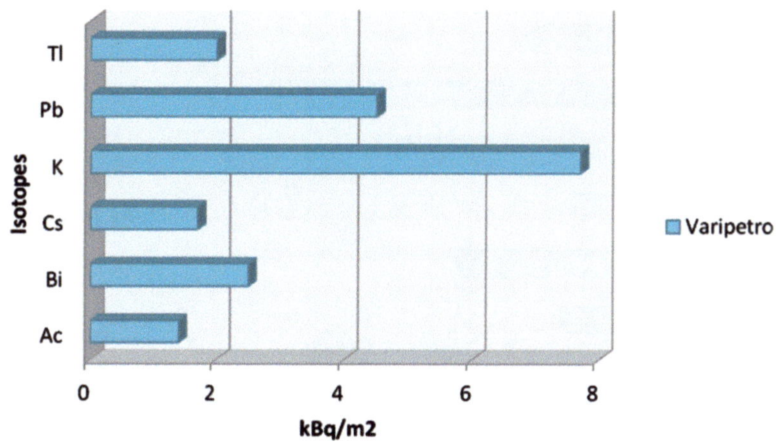

Fig. B.17 Radioactive concentrations in the prefecture of Xanion

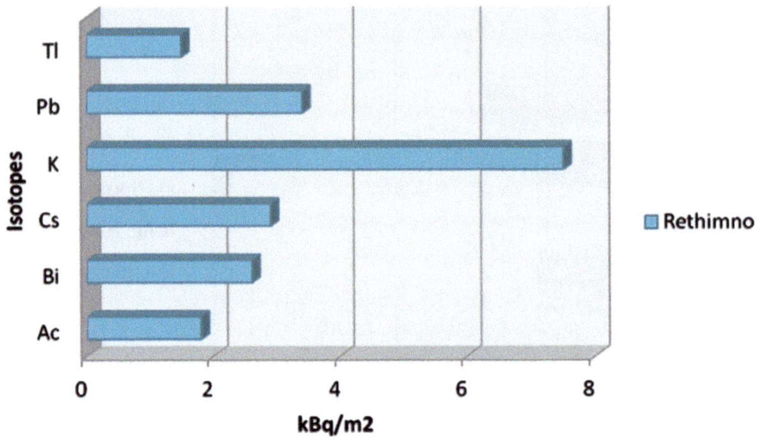

Fig. B.18 Radioactive concentrations in the prefecture of Rethimis

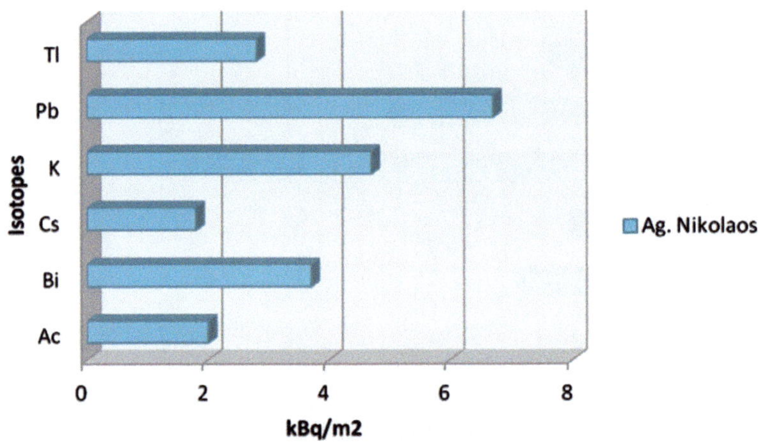

Fig. B.19 Radioactive concentrations in the prefecture of Lasithiou

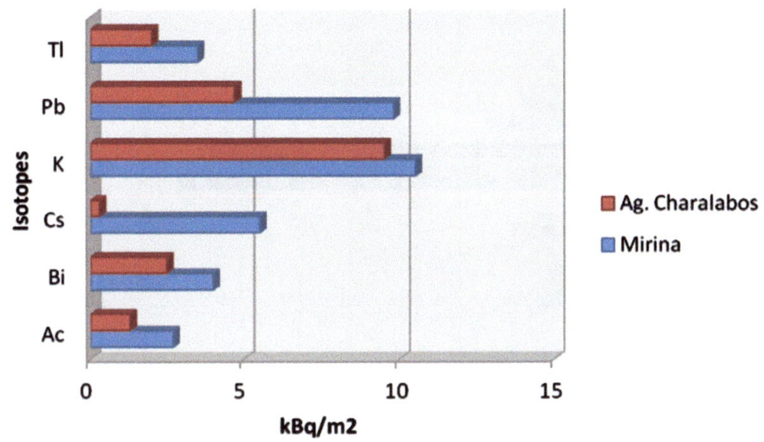

Fig. B.20 Radioactive concentrations in the prefecture of Lesvou

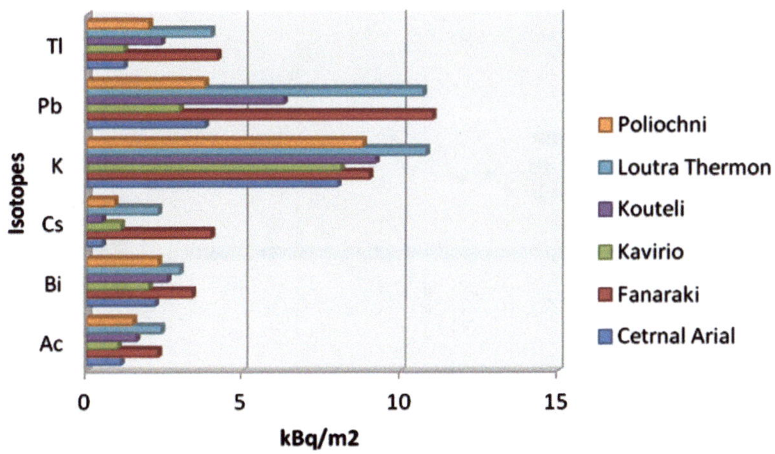

Fig. B.21 Radioactive concentrations in the prefecture of Samou

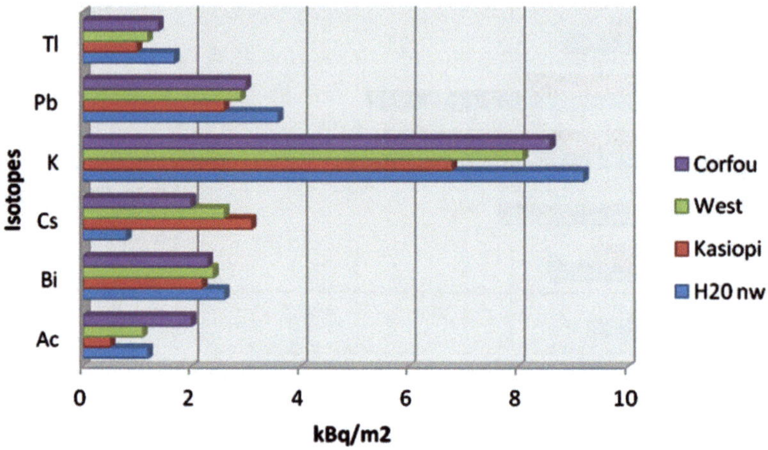

Fig. B.22 Radioactive concentrations in the prefecture of Kerkiras

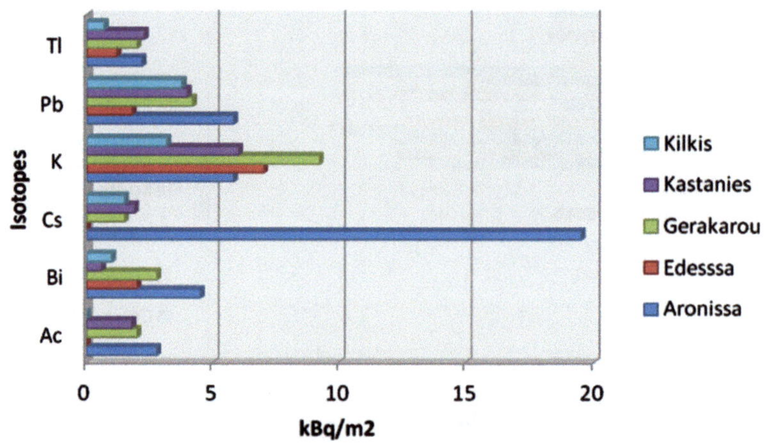

Fig. B.23 Radioactive concentrations in the prefecture of Kilkis

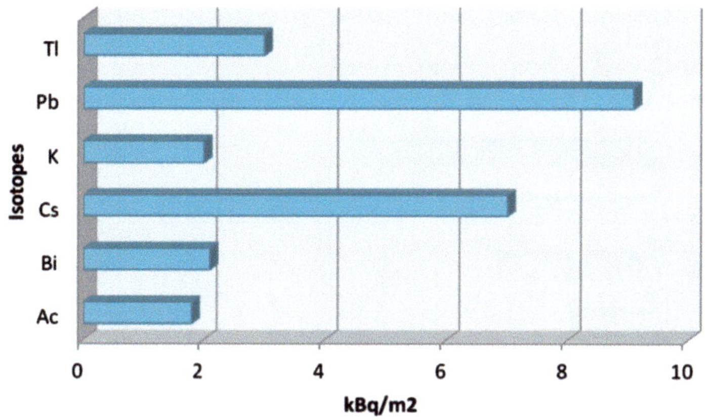

Fig. B.24 Radioactive concentrations of Holy Mountain (Athos)

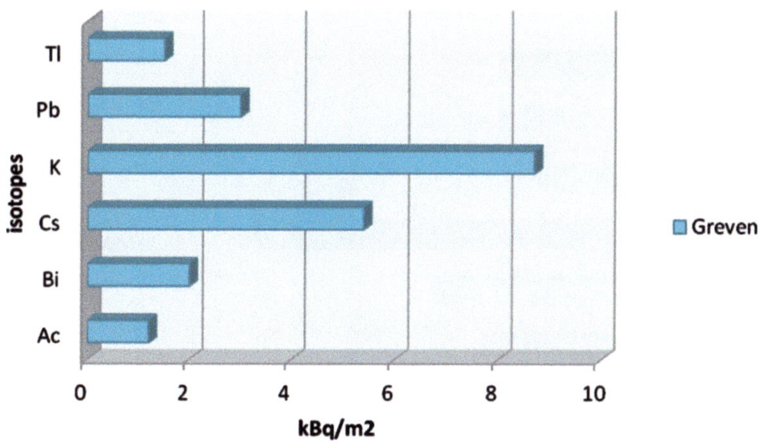

Fig. B.25 Radioactive concentrations in the prefecture of Grevenon

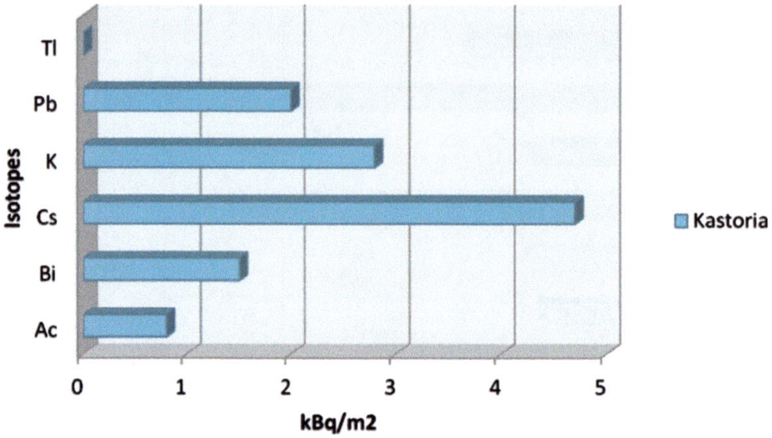

Fig. B.26 Radioactive concentrations in the prefecture of Kastorias

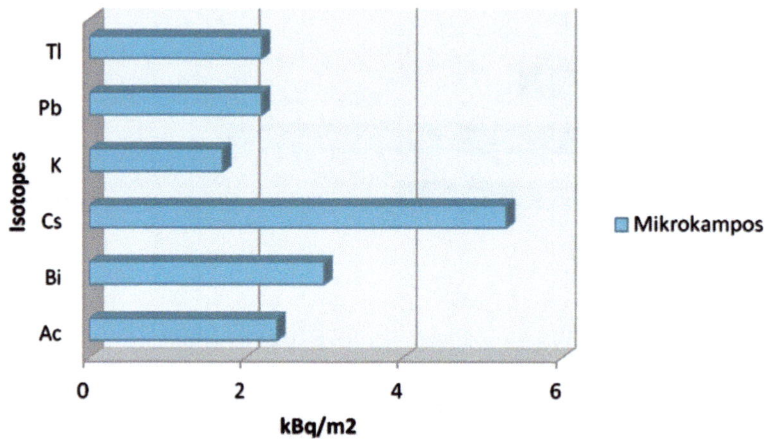

Fig. B.27 Radioactive concentrations in the prefecture of Dramas

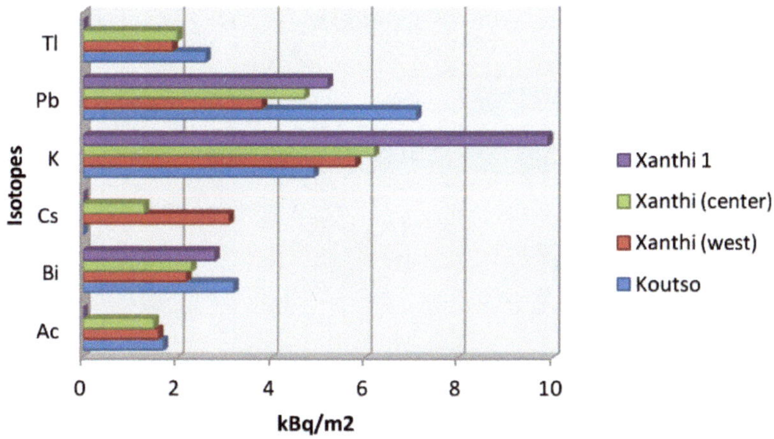

Fig. B.28 Radioactive concentrations in the prefecture of Xanthis

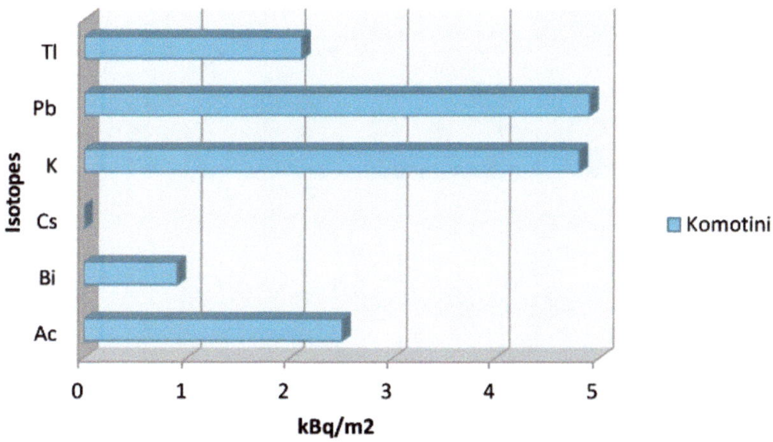

Fig. B.29 Radioactive concentrations in the prefecture of Rodopis

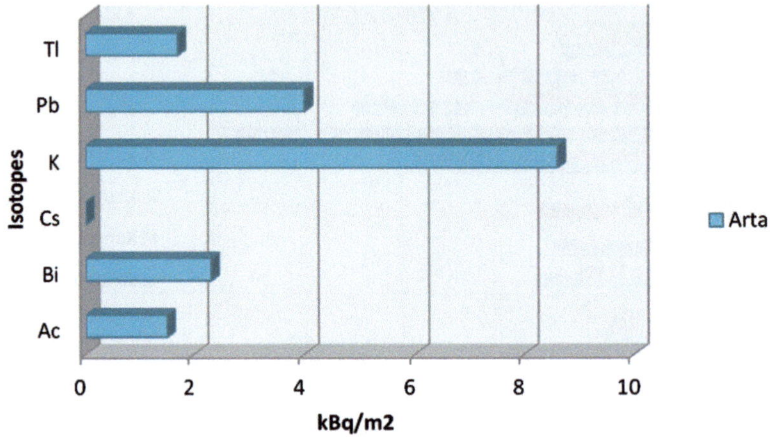

Fig. B.30 Radioactive concentrations in the prefecture of Artas

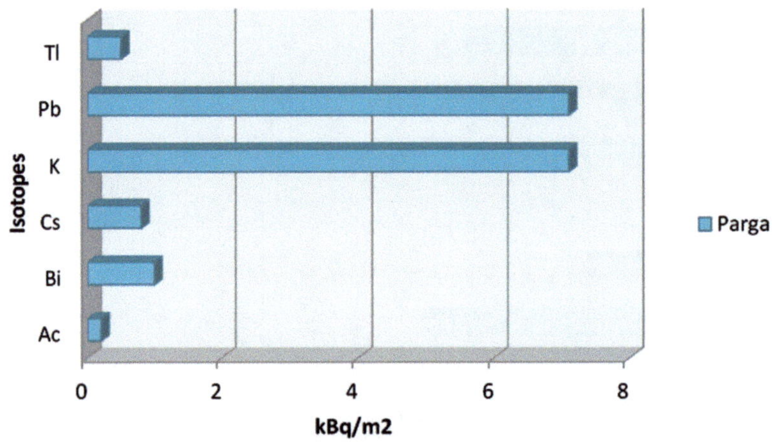

Fig. B.31 Radioactive concentrations in the prefecture of Prevezis

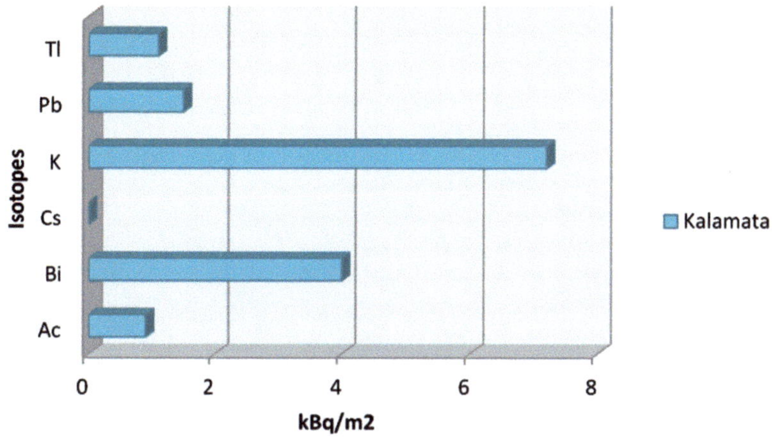

Fig. B.32 Radioactive concentrations in the prefecture of Messinias

References

1. A. Aarkrog et al., *Environmental Radioactivity in Denmark in 1986* (Riso National Laboratory, Denmark, 1988), p. 549
2. F. Adams, R. Dams, *Applied Gamma-Ray Spectrometry*, 2nd edn. (Pergamon Press, Oxford, 1970)
3. A.G. Kelesis, *The Atmospherical Pollution with SO_2 at the Area of Thessaloniki Prom* (Aristotles University of Thessaloniki, Thessaloniki, 1989)
4. Ph. Albert, *L'analyse par radioactivation, Applications des sciences nucleaires* (Gauthier-Villard & Cie, Paris, 1964)
5. American National Standards Institute, Calibration and Usage of Germanium Detectors for Measurement of Gamma-Ray Emission Rates of Radionuclides, Rep. No 42.14, ANSI, New York, 1978
6. American Society For Testing and Materials, *Standard Practice for High-Resolution Gamma-Ray Spectrometry of Water*, vol.11.02 (ASTM Standards, Philadelphia, 1986)
7. American Society For Testing and Materials, *Standard Practice for the Measurement of Radioactivity*, vol.11.02 (ASTM Standards, Philadelphia, 1986)
8. Anonymous, Measurements of Radionuclides in food and the Environment, Technical reports series No 295, International Atomic Energy Agency (IAEA), Vienna, 1989
9. Anonymous, International Dairy Federation (IDF), Standards 20/1962, 21/1962, 1/1963
10. M. Antonopoulos-Domis, A. Klouvas, F. Tervsidas, Deposition of the long-lived radionuclides ^{137}Cs and ^{134}Cs in the soils of Macedonia and Thrace due to Chernobyl accident. *Proceedings of the Symposium on Impact of the Chernobyl Accident in Greece*, Athens, 19–20 November 1987
11. M. Antonopoulos-Domis, Investigation of long term impact of ^{134}Cs and ^{137}Cs in the agricultural and dairy production in N. Greece, Thesis for final year assessment, University of Thessaloniki, 1990
12. APHA, in *Standard Methods For The Examination Of Dairy Products*, 14th edn. ed. by E.H. Mar (APHA, Washington, DC, 1978)
13. P.A. Assimakopoulos, K.G. Ioannides, A.A. Pakou, H.V. Papadopoulou, Distribution of the radioisotopes 131I, 134Cs and 137Cs from the fallout following the accident at the Chernobyl nuclear reactor in cheese and other cheese making hyproducts. J. Dairy Sci. **70**, 1338–1343 (1987)
14. Association Of Official Analytical Chemists, *Iodine-131, Barium-140 and Ceasium-137 in Milk and Other Foods—Gamma-ray spectroscopic method, Official Methods of Analysis*, 14th edn. ed. by S. Williams, 2nd Supplement (AOAC, Washington, DC, 1986), pp. 385–386

F. K. Vosniakos, *Radioactivity Transfer in Environment and Food*, Environmental Science and Engineering, DOI: 10.1007/978-3-642-28741-1,

15. A.W. Brewer, A replacement of the Dobson spectrophotometer. Pure App. Geophysics. **106–108**, 919–927 (1973)
16. F. Pasquill, *Atmospheric Diffusion*. (Wiley, New York, 1974)
17. U. Bear, D. Zavlasky, S. Irmay (eds.), *Physical Principles of Water Percolation and Seepage* (UNESCO, Paris, 1968)
18. K.E. Van Berveijk et al., J. Environ. Radioact. **15**, 265–276 (1992)
19. J.B. Birks, *Theory and Practice of Scintillation Counting* (Pergamon Press, London, 1964)
20. G. Bonazzola, R. Ropolo, A. Fachinelli, Profile and downward migration of ^{137}Cs and ^{106}Ru deposit on Italian soils after the Chernobyl accident. Physics **645**, 479–484 (1993)
21. L.Z. Bodnar, D.R. Percival (eds.), Anlytical Chemistry Branch Procedures Manual Rep. IDO-12096, Idaho National Engineering Laboratory, ID, 1982
22. F. Botchever, N. Lapchin, A. Oradovskaya, *Protection des eaux souterraines* (Nedra, Moskow (Russe), 1979)
23. British Standard 1741/1963
24. R.J. Budnitz, A.V. Nero, D.J. Murphy, R. Graven, *Instrumention for Environmental Monitoring*, vol. 1, 2nd edn. (Wiley, New York 1983)
25. C.S. Zerefos, N. Zoumakis, The Atmospherical Pollution with SO_2 of Thessaloniki Data (Ministry of Macedonia and Thrace, Thessaloniki, 1984)
26. J.-M. Cavedon, *La radioactivite* (Flammarion, Paris, 1996)
27. P. Cawse, S.J. Baker, A.S. Grandison, M.J. Lewis, S. Patel, The influence of processing on the radionuclide content of food, (Prog. Int. Sem., Caradache, 1991), Rep.EUR 14469, pp. 271–188
28. C. Chassard-Bouchaud, Environnement et radioactivite, No2797, PUF, “Que sais-je?”, 1993
29. G.R. Choppin, J. Rydberg, *Nuclear Chemistry and Applications* (Pergamon Press, Oxford, 1980)
30. G.B. Cook, J.F. Duncan, *Modern Radiochemical Practice* (Claredon Press, Oxford, 1958)
31. L. Costrell, D.E. Persyk, C. Sanderson, G. Walfford, F.J. Walter, Germanium semiconductor detector efficiency determination using a standard Marinelli beaker geometry. Health Phys. **38**, 229–232 (1980)
32. S. Danali, St. Papayannakis, Correlation between air radioactivity and air masses in Greece during the period 1964–1972. Health Phys. **28** 727–732 (1975)
33. S. Danali, St. Papaynnakis, Health Phys. **28**, 727 (1975)
34. S. Danali, Translation report No 7208, United States Atomic Energy Commission (1972)
35. Das Minimal-Prinzip, Information material of RWE, W. Germany
36. F. David, J.-P. Schapira, Le retraitement des combustibles nucleaires. Recherche **111**, 520 (1980)
37. J.G. Davis, T.R. Ashton, M. McCaskill, Dairy Ind. **36**, 569 (1971)
38. J.G. Davis, T.R. Ashton, M. McCaskill, Enumeration and viability of *L. bulgaricus* and *S. thermophilus*in yoghurts. Dairy Ind. **36**, 569–573 (1971)
39. K. Debertin, Messanleitung fur die Bestimmung von Gammastrahlen-Emissionsraten mit Germanium-Detectoren, Rep. PTB-Ra-12, Physikalisch-Technische Bundesanstelt, Brawnschweig (1980), 84 S
40. K. Debertin, U. Schotzic, Coincidence summing corrections in Ge (Li)—spectrometry at low source-to-detector distances. Nucl. Instrum. Methods **158**, 471–477 (1979)
41. K. Debertin, U. Schotzig, K.F. Walz, H.M. Weiss, Efficiency calibration of semiconductor spectrometers—Teechniques and accuracies (Proc. ERDA Symp. on X- and Gamma-Ray sources and applications). Rep. CONF-760539, National Technical Information Service, Springfield, 1976
42. M. De Bruin (ed.) Glossary of terms used in nuclear analytical chemistry. Pure Appl. Chem. **54**(8), 1533–1554 (1982)
43. G. Dechant, S. Donivan, Calibration and performance of a high-resolution Gamma-spectrometry system, Rep. GJBX-4-84 (Bendix Field Engineering Corp., Grand Junction, 1984)

44. D.H. Denham, D.A. Waite, J.P. Corley, Summary of selected AEC contractor environmental surveillance techniques and capabilities, Battelle, Pacific Northwest Laboratories Report BNWL-B-384, 1984
45. N.W. Desrosier, *The Technology at Food Preservation* (AV1 Publishing, Westport, 1963)
46. L. Devell, H. Tovedal, U. Bergstrom, A. Appelgren, J. Chyssier, L. Anderson, Initial observations of fallout from the reactor accident at Chernobyl. Nature **321**, 192–193 (1986)
47. K. Dorfner, *Ion-exchangers—Properties and Applications* (Ann Arbor Science, Ann Arbor, 1972)
48. H.J. Dunster, H. Howells, W.L. Templeton, District surveys following the Windscale incident (Prog. 2nd Int. Con. Geneva, 1985) pp. 296–308
49. AEA, Handbook of parameter values for the predition of radionuclide transfer in technical reports Serie No364, International Atomic Energy Temperate Environments Agency, Vienna, 1994
50. J.D. Eakins et al., Analytical procedures used by the bioassay section at AERE Harwell, Rep. AERE-AM-103, Harwell Laboratory, Oxford shire, 1968
51. F. Eekhof-Stork, *The World Atlas of Cheese* (Paddington Press, New York, 1976)
52. M. Eisenbud, *Environmental Radioactivity*, 2nd edn. (Academic Press, New York, 1973)
53. Environmental Surveillance Group H-8, Environmental surveillance at Los Alamos Scientific Laboratory Report LA-9762-ENV, 1983
54. Environmental Surveillance Group H-8, Environmental surveillance at Los Alamos during 1982, Los Alamos Scientific Laboratory Report LA-9762-ENV, 1983
55. M. Errera, Les effets des radiations nucleaires a faibles doses. La recherche **168**, 958–968 (1985)
56. R.D. Evans, *The Atomic Nucleus* (McGraw-Hill, New York, 1955)
57. A. Freese, J. Cherry, *Groundwater* (Prentice-Hall, Englewood Cliffs, 1979)
58. G. Friedlander, J.W. Kennedy, J.M. Miller, *Nuclear and Radiochemistry*, 3rd edn. (Wiley, New York, 1981)
59. M.J. Frissel, J. Koster, Soil-to-plant transfer factors of radionuclides, expected values and uncertainties, a summary of available data. Contribution prepared for the meeting of the IUR soil-to-plant workgroup, p. 24, Egham, 14–16 April, 1987
60. M. Galabov, Migration des pollutants dans les nappes aquifers intensite de pollution variable aux limites—La methode de protection des eaux souterraines (Nedra, Moskow (Russe), 1985)
61. M. Galabov, *Prognostic du transport convectif de masse et de chaleur en conditions complexes* vol. XXXII, part IV (Annual of Higher Institute of Mining and Geology, Sofia, 1986)
62. M. Galabov, P. Pentchev, *Solution des problemes hydrogeologiques pratiques a l'aide de computer* (Technika, Sofia, 1986)
63. G.E. Geoffrey, J.W. Poston, *Principles of Nuclear Radiation Detection* (Ann Arbor Science Publishers, Ann Arbor, 1980)
64. S.E. Gilliland, M.L. Speck, Antagonistic action of *L. acidophilus* toward intestinal and foodborn pathogens in associative cultures. J. Food Prot. **40**, 820–823 (1979)
65. S. Gmuca, I. Ribansky, Efficiency calibration of Ge (Li) detectors in close geometry. Nucl. Instrum. Methods **202**, 435–437 (1982)
66. V. Goloubav, A. Garibianz, *Processus heterogenes de la migration geochimique* (Nedra, Moskow (Russe), 1968)
67. Greek Atomic Energy Commision, The Chernobyl nuclear accident and its consequences to Greece DEMO 86/3G, 1986
68. Greek Atomic Energy Commision, The Chernobyl nuclear accident and its consequences to Greece DEMO 86/10G, 1986
69. M.L. Green, A.S. Grandison, in *Secondary (Nonenzymatic) Phase of Rennet Coagulation and Post—coagulation Phenomena*, ed. by P.F. Fox. Cheese: Chemistry, Physics and Microbiology. General Aspects, vol. 1 (Elsevier Applied Science, London, 1987) p.125

70. A. Gringaten, J.P. Sauty, Simulation des transfers de chaleur dans les aquifers. Bull BRGM, Fr. Sect II, No 1 (1975)
71. T.P. Guinee, P.F. Fox, in *Salt in Cheese: Physical, Chemical and Biological Aspects*, ed. by P.F. Fox. Cheese: Chemistry, Physics and Microbiology. General Aspects vol. 1 (Elsevier Applied Science, London, 1987) p. 251
72. H. Bonka, Strahlenexposition durch radioaktive Emissionen aus kerntechnischen Anlagen im Normalbetrieb (Verlag TUV Rheinland, Cologne, 1982)
73. M. Haissinsky, *Nuclear Chemistry and its Applications* (Addition-Wesley, Reading, 1964)
74. M. Haissinsky, J.P. Adloff, *Radiochemical Survey of the Elements* (Elsevier, Amsterdam, 1965)
75. D. Halliday, *Introductory Nuclear Physics*, 2nd edn. (Wiley, New York, 1955)
76. J.H. Head, Minimum detectable photo peak areas in Ge (Li) spectra. Nucl. Instrum. Methods **98**, 419 (1972)
77. R.L. Heath, Gamma-ray Spectrum catalogue, Ge (Li) and Si (Li) Spoctrometry, vol.2, US atomic energy commission, Rep.ANCR-1000-2, Idaho Falls, 1974
78. J.F. Horwood, Brief communications, 20th Inter. Dairy Congr. I.E. 375, (1978)
79. International Atomic Energy Agency, *Methods of Low-Level counting and spectrometry (Proc. Symp. Berlin (West), 1981)*, (IAEA, Vienna, 1981)
80. International Atomic Energy Agency, *Methodology of Radionuclides (Proc. Symp. Vienna, 1959)* (IAEA, Vienna, 1960)
81. International Atomic Energy Agency, *Reference Methods for Marine Radioactivity Studies II, Technical Reports Series No 118* (IAEA, Vienna, 1970)
82. International Atomic Energy Agency, *Reference Methods for Marine Radioactivity Studies II, Technical Reports Series No 169* (IAEA, Vienna, 1975)
83. International Commission of Radiation Units And Measurements, Measurement of Low-Level Radioactivity, Rep. No 22, ICRU, Washington, DC, 1972
84. International Atomic Energy Agency, *Methods of Low level counting and Spectrometry* (Proc. Symp. Berlin (West), 1981) (IAEA, Vienna, 1981)
85. International Commission of Radiation Units And Measurements, Measurements of Low-level Radioactivity, Rep. No22, ICRU Washington, DC, 1972
86. International Dairy Federation (IDF), Standards 20/1962, 21/1962, 1/1963
87. International Dairy Federation (IDF), Standards 17/1961, 20/1962, 21/1962, 1/1963, 3/1963, 4/1963, 27/1964
88. J.A. Garland, J.R. Branson, The mixing height and mass balance of SO2 in the atmosphere above Great Britain. Atmos. Environ. **9**, 643-659 (1976)
89. J.B. Kerr et al., *Intercomparison of the Dobson and Grating ozone spectrophotometer* (Atm. Envir. Serv. of Canada, Internal Report ARPD, 1976)
90. Kerr J.B. et al., Measurements of ozone with Brewer spectrophotometer, in *Proceedings of Ozone Symp*, Boulder, 1976, ed. by J. London
91. J.B. Kerr et al., The automated Brewer spectrophotometer, in *Proceedings of Quadr. Ozone Symp* ed. by C.S. Zerefos, A. Ghazi (Chaldiki, Greece, 1984)
92. R. Jedlovszky, A. Polacsek, A. Andai, Accurate evaluation of gamma-ray spectra for the measurement of gamma-ray emission probabilities. Int. J. Appl. Radiat. Isot. **34**(8), 1057–1065 (1983)
93. F.B. Johns, P.B. Hahn, D.J. Thome, E.W. Bretthauer (eds.), Isotopic analysis by gamma-ray spectra using Lithium-drifted Grmanium detectors, Radiochemical analytical procedures for analysis of environmental samples, Rep., Las Vegas, NV (1979). Ibid, Isotopic analysis by gamma-ray spectroscopy using Thallium-activated Sodium Iodide crystals
94. F.B. Johns, Handbook of Radiochemical Analytical Methods, Rep. EPA-680/4-75-001, National Environmental Research Center, Las Vegas, 1975
95. F.B. Johns, P.B. Hahn, D.J. Thome, E.W. Bretthauer (eds.), Radiochemical Analytical Procedures for Analysis of Environmental Samples, Rep.EMSL-LV-0539-17, Environmental Monitoring and Support Laboratory, Las Vegas, 1979

96. JRC-rem Project, Total ^{137}Cs Deposition from the Chernobyl accident: Validation Plot for Hellas, National Technical University Athens, 1986
97. C. Lewiner, Les centrales nucleaires, PUF "Que sais-je ?", No1037, (1991)
98. J.K. Kandarakis, E.M. Anifantakis, National Committee for Milk in Greece, vol.1 (1986) p. 20
99. G. Kanish, H. Ruhle, R. Schelenz (eds.), *Messanleitungen fur die Uberwachung der Raddioactivitat in der Umwelt* (Bundesminister des Inneren, Karsruhe, 1984)
100. J. Kleinberg, H.L. Smith (eds.), Collected Radiochemical Procedures, Rep. LA-1721, 4th edn., Los Alamos Scientific Laboratory, Los Alamos, 1982
101. G.F. Knoll, *Radiation Detection and Measurement* (Wiley, New York, 1979)
102. D.C. Kocher, *Radioactive Decay Data Tables* (Technical Information Center, Springfield, 1981)
103. F. Kosikowski, A.C. Vauddahlberg, A rapid direct—distillation method for determining the volatile fatty acids of cheese. J. Dairy Sci. **29**, 861–871 (1946)
104. F. Kosikowski, *Cheese and Fermented Milk Foods*, 2nd edn. 3rd printing, (F.V. Kosikowski and Associates, Brooktondalle, 1982) p. 535
105. H.L. Krieger, S. Gold, Procedures for Radiochemical Analysis of Nuclear Reactor Aqueous solutions, Rep. EPA-R473-014, Environmental Monitoring and Support Laboratory, Cincinati, 1980
106. H.L. Krieger, E.L. Whittaker, Prescribed Procedures for Measurement of Radioactivity in Drinking water, Rep. EPA-600/4-80-032, Environmental Monitoring and Support Laboratory, Cincinati, 1980
107. P. Kritidis, Kollas, Individual and social risk due to natural and artificial environmental radioactivity in Greece. Radiat. Prot. Dosim. **45**(1/4), 673–675 (1992)
108. P. Kritidis, E. Papanicolaou, Correlation between radioactive deposition and contamination of certain products with Caesium after the Chernobyl Nuclear Accident, in *Proceedings of Symp. On Impact of the Chernobyl Accident in Greece*, Athens, 19–20 November 1987
109. J. Krugers, *Instrumentation in Applied Nuclear Chemistry* (Plenum Press, New York, 1973)
110. R.E. Lapp, H.L. Andrews, *Nuclear Radiation Physics* (Prentice-Hall, New Jersey, 1964)
111. P.F. Larsen, M.C. Anon, Effect of water activity of milk upon growth and acid production by mixed cultures of *Streptococcus thermophilus* and *Lactobacillus bulgaricus*. J. Food Sci. **55**(3), 708–710 (1990)
112. H. Lauverier, The transport of heat in the oil layer caused by the injection of the fluid. Appl. Sci. Res. sect. A **5**, 145–150 (1955)
113. A.J. Lawrence, The determination of lactose in milk products. Aust. J. of Dairy Techn. **23**, 103 (1968)
114. C.M. Lederer, V.S. Shirley (eds.), *Table of Isotopes*, 7th edn. (Wiley, New York, 1978)
115. A.I. Lehniger, *Biochemistry* (World Publishers, New York, 1970) p. 133
116. K.C. Leung, S.Y. Lau, C.B. Poon, Gamma-radiotion dose from radionuclides in Hong Kong soil. Environ. Radioact. **11**, 279–290 (1990)
117. V. Lialko, M. Mitnik, Exploration des processus de transport de chaleur et de masse dans la croute terrestre (Naukova dunks, Kiev, 1978)
118. R. Lieberman (ed.), Radiochemistry Procedures Manual, Rep. EPA-520/5-84-006, Eastern Environmental Radiation Facility, Montgomery, 1984
119. Y. Lin, P. Lin, C. Chen, C. Huang, Measurements of terrestral gamma-radiation in Taiwan, Republic of China. Health Phys. **52**(6), 805–811 (1987)
120. L. Luckner, V. Schestakov, Modelisation de la migration des eaux souterraines. ed. "Nedra", Moskow (Russe) et VEB Deufcher Verlag fur grundstoffindustrie, Leipeig (Almand), 1986
121. E.H. Marth (ed.), *Standard Methods for the Examination of Dairy Products*, 14th edn. (American Public Health Association, Washington, DC, 1978)
122. G. Mavrogiaannopoulos, The inorganic nutrition of plants, Hydroponia, Book (Gr), 1992

123. G.J. McCallum, G.E. Coote, Influence of source-detector distance on relative intensity and angular correlation measurements with Ge(Li) spectrometers. Nucl. Instrum. Methods **130**, 189–197 (1975)
124. H.A.C. McKay, *Principles of Radiochemistry* (Butterworth, London, 1971)
125. F.A. Mettler, R.D. Moseley, *Direct Effects of Radiation*, in Medical Effects of Ionizing Radiation (Grune and Stration, New York, 1985)
126. P. Misaelidis et al., Distributor of fusion products in dust samples in Thessaloniki, in Greece Environment Pollution No47, 1987, pp.1–8
127. P. Misaelides, C. Sikalidis, R. Tsitouridou, C. Alexiadis, Distribution of fission products in Dust Samples from the Region of Thessaloniki, Greece, after the Chernobyl Nuclear Accident. Environ. Pollut. **47**, 1–8 (1987)
128. V. Mironenko, V. Rouminin, *Essaiad de migration dans les nappes aquifers* (Nedra, Moskow (Russe), 1986)
129. P. Misaelides, C. Sikalidis, R. Tsitouridou, C. Alexiadis, Radionuclide diffuse in the environment. Toxic. And Env. Chemistry **31**(32), 247–251 (1991)
130. G. Mocquot, R. Ducluzeau, I.D.F. Bull. **5**, 85 (1968)
131. G.K. Murthy, J.E. Campbell, Removal of radionuclides from milk. J. Dairy Technol. **47**(II), 88 (1964)
132. N. Zoumakis et al., The study of the changes of the total amount of SO_2, CO, NO_2 and O_3 at the area of Thessaloniki (in Greek), Internal report No 1/89 of T.E.I. (1989)
133. National Academy of Sciences, Nuclear science Series, NAS-NS, Office of Technical Services, Washington (1960)
134. National Council on Radiation Protection and Measurement, A Handbook of Radioactivity measurement procedures, Report No 58, NCRP, Washington (1978)
135. NCRP Reports, Environmental Radiation Mesurements, Washington (1988)
136. T.A. Nickerson, I.F. Vujicic, A.Y. Lin, J. Dairy Sci. **59**, 386 (1976)
137. A Ogata, R. Banks, A solution of the different equation of longitudinal dispersion in Porous Media. Gen. Survey Prof. Papers 411-1A (1961)
138. E.P. Papanikolaou, Radioactive fallout-plants and soil. Geoponica **208**, 224–229 (1972) (Gr)
139. E.P. Papanikolaou, P. Kritidis, Contamination of the agricultural land of Greece with radioactive Caesium and its effect on the growing crops. in Proceedings of international conference on environmental radioactivity in the Mediterranean area (Barcelona 10–13 May 1988), pp. 457–466
140. C.P. Papastefanou, M. Manolopoulou, S. Charalambous, Ceasium-137 in soils and its soil-to-plant tansfer rate, in Impact des accidents d'origine nucleaire sur l'environnement, I, Centre d'etudes nucleaires de Cadarache, 14–18 March 1988
141. B.S. Pasternack, N.H. Harley, Detection limits for radionuclides in the analysis of multi-component gamma-spectrometer data. Nucl. Instrum. Methods **91**, 533–540 (1971)
142. W.J. Price, *Nuclear Radiation Detection*, 2nd edn. (McGraw-Hill, New York, 1964)
143. M. Probonas, P. Kritidis, The exposure of a Greek population to natural gamma radiation of terrestral origin. Radiation Protection Dosim. **46**(2), 123–126 (1993a)
144. S.R. Pulusani, D.R. Rao, G.R. Sunki, Antimicrobial activity of lactic cultures: partial purifiction of antimicrobial compound(s) produced by S. thermophilus. J. Food Sci. **44**, 575–578 (1979)
145. P. Quittner, *Gamma-ray Speectroscopy* (Adam Higher, London, 1972)
146. P. Radvanyi, Les rayonnements nucleaires. PUF "Que sais-je?", No 844, (1995)
147. D.R. Rao, B.M. Reddy, G.R. Sunki, S.R. Pulusani, J. Food Quality **4**, 247 (1981)
148. D.R. Rao, B.M. Reddy, G.R. Smiki, S.R. Pulusani, Influence of methanol/acetone extract from thermophilus milk on selflife of ground beef steaks and milk. J. Food Quality **4**, 247–251 (1981)
149. M. Rapp, Milchwissenschaft **24**, 208 (1969)
150. M. Rapp, Uber das Eiweissabbvaurermogen van Milchaurebacterium. Milchwissenchaft **24**, 208–211 (1979)

151. J. Rasic, J.A. Kurmann, Yoghurt-Scientific Grounds, Technology. Manufacture and Preparations. vol. 1 (Technical Dairy Publishing House, Copenhagen, Denmark, 1978)
152. P. Reuss, L'energie nucleaire, PUF "Que sais-je?", No 317, (1994)
153. P. Robert, Dictionnaire alphabetique et analogique de la langue francaise, Paris, (1964)
154. C.G. Sanderson, An evaluation of commercial IBM PC software for the analysis of low level environmental gamma-ray spectra. in Proceedings of conference on low-Level techniques group—International committee of radionuclide metrology, (Wurenlingen, Switzerland 1987)
155. K. Savidou, C. Raptis, P. Kritidis, *Natural radioactivity and radon exhalation from building materials in Greece. Radiation Protection Dosimetry* (United Nations, New York, 1995)
156. G. Schaller, E. Wirth, C. Leising, Cs and K uptake by plants from Bavarian soils, in Impact des accidents d'origine nucleaire sur l'environnement, I, Centre d'etudes nucleaires de Cadarache, (14–18 March 1988)
157. R. Schelenz, E. Fischyer, Neutronenaktivierungsanalyse von Spurenelementen in Lebensmitteln, Rep. BFE 1976/4, Bundesforschungsanstalt fur Ernahrung, (Karlsruhe, 1976)
158. F.J. Schima, D.D. Hoppes, Tables for cascade-summing correlations in gamma-ray spectrometry. Int. J. Appl. Radiat. Isot. **34**, 1109–1114 (1983)
159. K. Siegbahn (ed.), *Alpha-Beta- and Gamma-Ray Spectroscopy* (North Holland, Amsterdam, 1965)
160. K. Sikalidis, P. Misaelidis, C. Alexiades, Caesium selectivity and fixation by vermiculite in the presence of various competing cations. Envir. Pollution **582**, 67–69 (1988)
161. S.E. Simopoulos, Soil sampling and ^{137}Cs analysis of the chernobyl fallout in Greece. Appl. Radiat. Isot. **40**(7), 65–69 (1989)
162. A. Sismanidou, D. Sapountzis, Electrical and electronic components of a diffuse radioactivity measuring station, Dissertation report, Department of Electrical Engineering, (Technological Educational Institute of Thessaloniki, 1985)
163. D. Soltner, *Les bases de la production vegetale, I: Le sol* (Sciences et techniques agricoles, Sainte-Gemmes-sur-Loire, 1975)
164. T.K. Taipale (ed.), *The Sampling and Analyzing Methods of Radionuclides Used in the Nordic Countries for Environmental Samples* (Rep. Nordic Liaison Committee for Atomic Energy, Stockholm, 1985)
165. A.Y. Tamine, H.C. Deeth, Yoghurt: Technology and biochemistry. J. Food. Prot. **43**, 939–977 (1980)
166. A.Y. Tamine, R.K. Robinson, Some aspects of the production of a concentrated yoghurt (labneh) popular in the Middle East. Milchwissenchaft **33**, 209–212 (1978)
167. D. Tang, E. Frind, E. Sudicky, Contaminant transport in fratured porous media: analytical solution for a sing fracture. Water Res. Res. **17**(3), (1981)
168. Th.-A. Batalas, F. Vosniakos, Radionuclide diffuse in the atmosphere (in Greek), Internal report No 1/89 (Laboratory of applied physics, Science Department, Technological Educational Institution of Thessaloniki (T.E.I.))
169. M. Tubiana, M. Bertin, Radiobiologie-Radioprotection, PUF "Que sais-je?", No 2439, (1989)
170. J.R. Terrill, J.G., jr., Public health radiation surveillance. Health Phys. **11**, 917 (1965)
171. M. Tubiana, R. Dautray, La radioactivite et ses applications, PUF "Que sais-je?", No 33, (1996)
172. M. Tubiana, J. Dutreix, A. Wambersie, *Radiobiologie* (Hermann, Paris, 1986)
173. United Nations Scientific Committee on the Effects of Atomic Radiation, *Report of the General Assembly* (United Nations, New York, 1977)
174. K.E. Van Bergeijk, H. Noordijk, J. Lembrechts, M.J. Frissel, Influence of pH, soil type and organic matter content on soil-to-plant transfer of Radioceasium and Strontium as analysed by a nonparametric method. in Radioactivity, J. Environ, (England, 1992), pp. 265–276
175. C.M. Vandecasteele, Th. Zeevaert, R. Kirchmann, Transfer des radionucleides dans des produits de l'agriculture, in Annales de l'Association belge de Radioprotection, vol. 20(2), (1995)
176. C.M. Vandecasteele, *Radioecologie* (Universite de botanique, Liege, 1998)

177. B.C. Veionglou, E.S. Boyazoglou, E.D. Kotouza, Dairy Indust Int. **44**(10), 29, (1979)
178. N. Verigun, B. Scherjukov Diffusion et echange masse lors de filtration des liquides en milieu poreux. Developpement des recherches de filtration en URSS. Ed. "Nedra, (Moskow, Russe, 1969)
179. N. Verigun et al., Methodes prognostiques de la salinite et sols et des eaux souterraines. Ed. "Kolos", (Moskow, Russe, 1979)
180. H.L. Volchok, G. DE PLANQUE, (eds.), Limit of Detection, EML Procedures Manual, Report HALS-300, 26th edn, (Environmental Measurement Laboratory, New York, 1983) D-08-05
181. F.K. Vosniakos, N.M. Zoumakis, C. Diomou, Cs-137 concentration in Greek soil due to Chernobyl accident. Fresenius Environ Bulletin **7**, 635–640 (1996)
182. F.K. Vosniakos, K.S. Farmakis, Diffuse radioactivity measurements and radionuclide identification in the air of the industrial area of Sindos—Thessaloniki before and after the arrival of the Chernobyl cloud. Toxicological Environ Chem **20–21**, 65–69 (1989)
183. F.K. Vosniakos, K.S. Farmakis, The radioactive contamination in Greece due to the Chernobyl accident and its impact to the environment. J. Environ. Prot. Ecol. **1**, 1–9 (2000)
184. IAEA. International Atomic Energy Agency, Technical Report Series No 295. Measurements of Radionuclides in food and the Environment, (1989)
185. H. Mundschenk, Zur Selectivitat und Empfindlichkeit von NAJ (TI) – UND Ge (Li)—Messonden in der radioekologichen und radiometrischen Wasseranalytik, Dtsch. Gewasserkd. Mitt. **20**, 13–20 (1976) (Teil I), Ibid,**22**, 161–173 (1978) (Teil II), Ibid, **23**, 97–106 (1979) (Teil III)
186. F.K. Vosniakos, A. Moumtzis, A. Kesidou, S. Ganatsios, A. Bizopoulos, P.A. Karakoltsidis, Transfer of ^{131}I and ^{137}Cs from cow milk to cheese and other products. Aust. J. Dairy Technol. **44**, 44–46 (1989)
187. F.K. Vosniakos, A. Moumtzis, A. Kesidou, T.H.A. Batalas, P.A. Karakoltsidis, Iodine-131 transport mechanism from feta cheese, upon immersion in water and neutral salt solutions. Aust. J. of Dairy Technol. **47**, 14–17 (1992)
188. F.K. Vosniakos, A. Moumtzis, A. Kesidou, P.A. Karakoltsidis, ^{131}I and ^{137}Cs removal from cheese upon immersion in a sodium chloride solution. Toxicol. And Environm. Chem. Vols. **31–32**, 241–246 (1991)
189. F. Vosniakos, A. Moumtzis, A. Giouvanoudi, G. Drosos, P. Karakoltsidis, Survival of lactic acid bacteria of yoghurt in the presence of ^{131}I. Dairy, Food Environ. Sanitation **11**(8), 433–435 (1991)
190. F.K. Vosniakos, N. Zoumakis, A. Giouvanoudi, C. Tsongas, P. Karakoltsidis, Radionuclide identification the air of the major area of Thessaloniki. MESAEP, 8th International Symposium, (Oct. 12–15, 1993, Antibes, France, 1993)
191. F. Vosniakos, T.A. Batalas, N. Zoumakis, A. Bizopoulos, Radionuclide diffuse in the environment. Toxic. Env. Chem. **31**(32), 247–251 (1991)
192. G. Weinberg, R. Williams, La solution nucleaire, in Pour la science, No 157, November (1990), p. 110
193. M.H. Weinberger, F.C. Luft, R. Block, D.P. Henny, J.H. Pratt, A.E. Weyman, L.T. Rankin, R.H. Murray, L.R. Willis, C.E. Grim, The blood pressure effect of high dietary sodium. Racial differences and the role of potassium. J. Amer. Coll. Nutr. **1**, 139–148 (1982)
194. L.G. Wilson, R.C., Bottomley, P.M. Sutton, C.H. Sisk, J. Soc. Dairy Technol. **41**, 10 (198)
195. World Health Organisation, *Methods of Radiochemical Analysis* (WHO-FAO-IAEA, Geneva, 1966)
196. S. Fetter, K. Tsipis, Les differents perils nucleaires, in Pour la science, No 63, Janvier (1983), p. 92
197. F. Vosniakos, Decontamination of milk from the radioisotopes iodium-131 and cesium-134,137. Adv. Food Sci. (CMTL) **18**(1–2), 52–55 (1996)
198. F. Vosniakos, K.A. Hatziioannou, The preservation and maturing of diary products by using an irradiation process. Adv. Food Sci (CMTL) **19**(1–2), 8–11 (1997)

199. F. Vosniakos, K. Zavlaris, T. Papaliaga, K. Vosniakos, Investigation on natural radioactivity concentration in building materials and the impact to human health. MASAEP—13th Symposium, Thessaloniki, 8–12 October 2005
200. Krishnamurthy, D.R. Bongirwar, Design consideration for food irradiators in developing countries. Food Irradiation Precesing, IAEA, Vienna, STI/PUB/695, ISBN 92-0-010185-2, (1985) pp. 353–363
201. J.G. Leemhorst, Industrial application of food irradiation, Food irradiation Now. in Proceedings of a symposium in Ede, (The Netherlands, Martinus Nijhoff/De.W.Junk Publishers, The Jague, 1982), pp. 60–68
202. P. Loaharanu, Asia regional cooperative project on food irradiation (RPFI), Food Irradiation Processing. in Proceedings of a symposium, (Washington, 1985), pp. 175–184, 1985
203. J. Masefield, G.R. Dietz, W.M. Owens, Major considerations governing the adaption of cobalt-60 sterilization. Radia. Phys. Chem. **14**, 277–283 (1979)
204. J. McKeown, N.K. Sherman, Linac based irradiatiors. Radiat. Phys. Chem. **25**, 103–109 (1985)
205. M.L. Moinet, *Aliments Irradies Le Fraicheur qui dure* (Science et Vie, Dec, 1983)
206. J. Schwartz, B. Hogeman, C. Hutting, R. Kellerman, W. Staege, B. Methodenbuch, and VI, Editor Neumann - Verlag, (Radebeul Berlin, 1950)
207. V. Stenger, T. Sipos, L. Laszlo, O. Hargittai, A. Kovacks, I. Harvarth, Experiences with a high capacity industrial scale pilot onion irradiator. Radiat. Phys. Chem. **22**, 717–732 (1983)
208. N. Tamura, *Gamma irradiation facilities* (Proceedings of the International Meeting on Radiation Processing, Tokyo, Japan, 1980), pp. 273–283
209. F. Vosniakos, K. Zavlaris, T. Papaliagas, Indoor concentration of natural radioactivity and the impact to human health. J. Environ. Protect. Ecology–JEPE**4**(3), 733–737 (2003)
210. T.S. Park, W.J. Jeon, Radio. Nucl. Chem (Articles) **193**(1), 133 (1995)
211. J. Beretka, P. Mathew, Health Phys. **48**, 87 (1985)
212. I. Noorddin, Determination of natural activity in building materials by direct gamma spectroscopy. Fresenius Envir. Bull. **8**, 027–077 (1999)
213. H.E. Jones, J.R. Cunningham, The Physics of Radiology. (Charles C Thimas, USA, 1983), p. 536
214. G.F. Knoll, *Radiation detection and measurements* (John Wiley & Sons, USA, 1989), p. 60
215. K. Papastefanou *Radiation Physics and Radioisotope Applications* (Zitis, Greece, (1992), p. 223
216. National Radiological Protection Board (NRPB). Board Statements on radon in homes. Documents of the NRPB**I**(1), (1990)
217. P. Ayotte, B. Levesque, D. Gauvin, Indoor exposure to ^{222}Rn. A public health perspective. Health Phys. **75**(3), 297–302 (1998)
218. F.T. Najafi, Radon reduction systems in the construction of new houses in Gainsvite, Florida. Health Phys. **75**(5), 514–517 (1998)
219. Greek Atomic Energy Commission 1986. The Chernobyl Nuclear Accident and Its Consequences to Greece. DEMO 86/3G, DEMO 86/10G
220. E.P. Papanikolaoy, Radioactive Fallout—Plant and Soil. Geoponia **208**, 224 (1972) (Greek)
221. K. Sikalidis, P. Misaelidis, K. Alexiadis, Cs Selectivity and fixation by vermiculate in the presence of various competing cations. Env. Pollution **52**, 67 (1998)
222. P. Kritidis, G. Kollas, Individual and social risk due to natural and artificial environmental radioactively in Greece. Rad. Prot. Dosimetry **45**(1–4), 673 (1992)
223. D. Antonopoulos, Long Term Effect of Cs-137 on Agricultural and Diary Production of Northern Greece, PhD Thesis, (Dept. of Electrical Engineering and Agronomy, University of Thessaloniki, Greece, 1990)
224. E. Papanikolaoy, P, Kritidis, Contamination of the Agricultural Land of Greece with Cs-137 and Its Effects on Crops. in Proceedings of international conference on environmental radioactivity in the Mediterranean Area, (Barcelona, May 1988), pp. 457–466

225. F.K. Vosniakos, K.S. Farmakis, The radioactive contamination in Greece due to Chernobyl accident and its impact to the environment. J. Environ. Protect. Ecol.**1**(1), (2000) pp. 1–9
226. K.S. Farmakis, F.K. Vosniakos, B. Koninckx, Soil to plant corn transfer of Cs 137 AND K-40 in Northern Greece. Fresenius Environ. Bulletin **9**, 623–630 (2000)
227. F. Vosniakos, K. Zavlaris, C. Mallas, D. Ivanova, Natural radioactivity concentration in building materials in Greece. J. Environ. Protect. Ecol. – JEPE**2**(1), 29–33 (2001)
228. H. Florou, P. Kritidis, F. Vosniakos, G.G. Polikarpov, V. Egorov, R. Delfanti, C. Papucci, A. Aladjadjiyan, Dispersion of 137Cs in the Eastern Mediterranean and the Black Sea: The Time Evolution in Relation to the Sources and Pathways. J. Environ. Protect. Ecol. – JEPE, **3**(1), 30–36 (2002)
229. M. Katsantoni, K. Penkyovska, K. Kazashka, J. Triandafyllis, F. Vosniakos, A. Aladjadjiyan, Measurements of Car exhaust gases and noise from automobiles in the major area of Kozani. J. Environ. Protect. Ecol.–JEPE **3**(1), 53–60 (2002)
230. H. Florou, P. Kritidis, F. Vosniakos, J. Triandafyllis, R. Delfanti, C. Papucci, A. Cigna, G.G. Polikarpov, V.N. Egorov, A.S. Bologa, V. Patrascu, Ceasium-137 in the Eastern Mediterranean—Impact Sources and Marine Pathways. Fresenius Environ. Bulletin **12**(1), 3–9 (2003)
231. D. Savidis, V. Grammatikis, J. Triandafyllis, F. Vosniakos, N. Zoumakis, A. Kelesis, Four-stroke diesel engines run on vegetable oil Mixtures—Performance and environmental effects. Fresenius Environ. Bulletin **15**(8b), 898–904 (2006)
232. E. Pantziarela, P. Papapreponis, I. Tsikopoulos, O. Serafimidou, F. Vosniakos, E. Vagtadli, The after-effects of atmospheric pollution on the first line blood assays. XXXIII World Congress of the International Society of Hematology, (Jerusalem, Israel, October 10–13, 2010)
233. H. Florou, P. Kritidis, Gamma radiation measurements and dose rate in the coastal areas of a volcanic island, Aegean Sea. Greece Rad. Protect. Dosim. **45**(1/4), 277–279 (1992)
234. P. Kritidis, H. Florou, *Natural Radioactivity in the Environment and Radioactivity Pollution* (National Conference on Environmental Sciences and Technology, Aegean University, Mytilini, Proceedings, vol .B, 1989), pp. 24–34
235. N. Ibrahim, Concentration of U and Th in Langkawi Black S and using INAA. Fresenius Environ. Bulletin **4**, 685 (1995)
236. N. Ibrahim, Determination of natural radioactivity in building materials by direct gamma spectroscopy. Fresenius Environ. Bulletin **8**, 72 (1999)
237. Radiological risk assessment in the S.E. Mediterranean Area. F. Vosniakos, A. Cigna, G. Vasilikiotis, P. Foster (eds.) (vol. I, II) (Technological Educational Institution of Thessaloniki, Greece, 2000)
238. M. Probonas, P. Kritidis, The exposure of the Greek population to natural gamma radiation of terrestrial origin. Radiol. Protect. Dosim. **46**(2), 123–126 (1993b)
239. A.C. Paul, F.M.D. Pillai, T. Velayudhan, K.C. Pillai, Internal exposure at high background areas. Natural radiation environment. in Proceedings of the international symposium on natural radiation environment. (Wiley Eastern Ltd. BARC, Bombay, India, 1982) pp. 50–57
240. A.P. Radhakrishna, H.M. Somashekarappa, Y. Narayan, K. Siddapa, A new natural background radiation area on the Southwest Coast of India. Health Phys. **65**(4), 390–395 (1993)
241. Society of Nuclear Medicine, Inc. Low Level Radiation Effects. (Society of Nuclear Medicine, New York, NY, 1982)
242. I. Noorddin, M.R. Yusof, M. Siphon, Environmental radioactivity level from the incineration of oil-sludge. Fresenius Environ. Bulletin **18**(6), 1025–1028 (2009)
243. A. Hoseyin, B. Ridvan, Radioactivity measurements and dose assessment from surface soils around the catalogzi coal-fired power plant, Turkey. Fresenius Environ. Bulletin **20**(7), 1642–1648 (2011)
244. L. Misheva, Radiochemical determination of plutonium in soil samples. J. Environ. Protect. Ecol.–JEPE **5**(3), 687–691 (2004)
245. M. Banov, Survey of lands affected by uranium mining, their reclamation and use in Republic of Bulgaria. J. Environ. Protect. Ecol–JEPE **5**(3), 692–697 (2004)

246. M.M. Brasoveanu, M.N. Grecu, M.R. Nemtanu, A. Tirlea, R. Minea, Irradiated dried fruits identification. J. Environ. Protect. Ecol.–JEPE **5**(4), 917–921 (2004)
247. M.R. Nemtanu, A. Tirlea, M. Brasoveanu, Physical methods for irradiated food identification. J. Environ. Protect. Ecol.–JEPE **5**(4), 922–925 (2004)
248. I. Bolocan-Viasu, Radiation exposure from radon-222 and its daughters. J. Environ. Protect. Ecol.–JEPE **6**(2), 437–440 (2005)
249. D. Todorovic, D. Popovic, M. Radenkovic, Radionuclides in ground level air in Belgrade urban area. J. Environ. Protect. Ecol.–JEPE **6**(4), 956–960 (2005)
250. TZ. Bineva, D. Staneva, P. Yonova, I. Yordanova, N. Stoeva, Modification of the irradiation effect on oat plant during the vegetation period. J. Environ. Protect. Ecol.–JEPE **7**(1), 146–150 (2006)
251. I. Yordanova, Natural radioactivity in samples from some regions in the Rila mountains and around former uranium mining areas. J. Environ. Protect. Ecol.–JEPE **7**(1), 151–156 (2006)
252. L. Misheva, Methodological features for determination of plutonium isotopes in soil samples. J. Environ. Protect. Ecol.–JEPE **7**(1), 157–161 (2006)
253. C. Varlam, I. Stefanescu, A. Feru, I. Popescu, A. Enache, M. Raceanu, M. Varlam, Tritium measurement using direct liquid scintillation counting in environmental aqueous samples. J. Environ. Protect. Ecol.–JEPE **7**(1), 162–169 (2006)
254. I. Yordanova, D. Staneva, TZ. Bineva, Radioactivity measurements in the region of Belene along the Danube river. J. Environ. Protect. Ecol.–JEPE **7**(1), 170–175 (2006)
255. M. Zucchetti, F. Aumento, Accidents in nuclear-powered submarines and their effects on environmental marine pollution. J. Environ. Protect. Ecol.–JEPE **7**(1), 176–185 (2006)
256. T.Z. Bineva,, D. Staneva, N. Stoeva, I. Yordanova, Effect of the synthetic growth regulators on gamma-irradiation stress in triticale plants. J. Environ. Protect. Ecol.–JEPE **7**(3), 675–678 (2006)
257. M. Zucchetti, Doses due to contamination of the Techa river by the Mayak nuclear complex in Russia. J. Environ. Protect. Ecol.–JEPE **7**(4), 918–925 (2006)
258. C. Bucur, C. Olteanu, M. Pavelescu, Sorption of cesium-137 on soil samples. J. Environ. Protect. Ecol. JEPE**8**(3), 649–654 (2007)
259. M. Toma, L. Dinescu, I. Paunica, C. Olteanu, Study of the radioactivity level around the National Radioactive Waste Repository, Baita –Bihor. J. Environ. Protect. Ecol.–JEPE **8**(4), 919–925 (2007)
260. N. Mocanu, A. Ranca, N. Paunescu, G. Marin, Environmental tritium related to normal operational stage of the candu nuclear power plant at Cernavoda—Romania. J. Environ. Protec. Ecol.–JEPE **8**(4), 926–933 (2007)
261. I. Yordanova, D. Staneva, A. Zlatev, L. Misheva, Tz. Bineva, M. Poinarova, Study of the radioceasium content in Bulgarian mushrooms for the year of 2005. J. Environ. Protect. Ecol.–JEPE**8**(4), 934–939 (2007)
262. L. Bonavigo, M. Zucchetti, Trino vercellese nuclear power plant decommissioning. J. Environ. Protect. Ecol.—JEPE**9**(3), 665–674 (2008)
263. R. Margineanu, N. Popa, C. Olteanu, ^{137}Cs Use in sedimentation rate assessment in gullies and reservoirs in Romania. J. Environ. Protect. Ecol.—JEPE**8**(1), 197–203 (2007)
264. C. Cosma, I. Petrescu, C. Meilescu, A. Timar, Studies on the radioactivity of lignite from the area between the Danube and Motru (South-West Romania) and the incidence on the environment. J. Environ. Protect. Ecol.—JEPE **10**(1), 192–200 (2009)
265. A.A. Cigna, Impact of the Chernobyl accident in Western Europe. J. Environ. Protect. Ecol.—JEPE**10**(2), 476–487 (2009)
266. A.A. Cigna, Anthropogenic radionuclides in surface sea water and fallout. J. Environ. Protect. Ecol.—JEPE **11**(2), 666–674 (2010)
267. I. Tanaskovic, D. Golobocanin, S. K. Petrovic, N. Miljevic, Natural radioactivity in Serbian Spa Waters. J. Environ. Protect. Ecol.—JEPE**11**(1), 286–294 (2011)
268. F. Cfarku, E. Bylyku, G. Xhixha, M. Shyti, E. Bakiu, Radioactivity monitoring of drinking water of Albania. J. Environ. Protect. Ecol.–JEPE**12**(3), 1116–1120 (2011)

269. A. Preda, D. Costinel, C. Mladin, A. Miu, C. Barbu, S. Pintilie, S. Sbirna, Analysis of isotopic ratio of hydrogen from the Olt river by mass-spectrometry. J. Environ. Protect. Ecol—JEPE**11**(2), 1458–1464 (2010)
270. International Atomic Energy Agency, Measurement of radionuclides in food and the environment, Technical Report Series No. 295 (IAEA—Vienna, 1989)
271. R. Bojanowsk, S. Ballestra, R. Fukai, Natural radionuclides in the North-Western Mediterranean Sea. Rapp. Comm. Int. Mer. Medit. **29**, 725 (1982)
272. Y. Miyake, M. Mayeda, Y. Sugimura, Uranium content and the activity ratio $^{234}U/^{238}U$ in marine organism and seawater in the Western North Pacific. J. Oceanogra. Soc. Jpn. **26**, 123 (1970)
273. W.S. Broecker, Y. Li, Cromwell, radium-226 and radon-222: concentrations in Atlantic and Pacific Oceans. Science **158**, 1307 (1967)
274. B.J. Szabo, Radium content in pland seawater. Geochim. Cosmochim. Acta **32**, 1321 (1967)
275. W.S. Moore, W.M. Sackett, Uranium and thorium series in equilibrium in seawater. J. Geophys. Res. **69**, 540 (1964)
276. A. Kaufman, Uranium and thorium series isotopes programme. USAEC Rep. CU-3139-1
277. B.L.K. Somayajulu, E.D. Goldberg, Thorium and uranium isotopes in seawater and sediments. Earth Planet Sci. Lett. **1**, 102 (1966)
278. A. Cigna, R. Delfanti, R. Serro, The radiological exposure of the population of the European Community to radioactivity in the Mediterranean Sea. Marina-Med Project. Radiation Protection 70, Report EUR I5564 EN (1994)
279. J.M. Willemenot, R. Schnepf, G. Fraser, Radioactive effluents from nuclear power stations and nuclear fuel reprocessing plants in the European Community, 1991–1995. Radiation Protection 104. ISSN 1018-5593 (1999)
280. F.K. Vosniakos, P.A. Karakoltsidis, A. Kalfa, A. Giouvanoudi, A. Moumtzis, Pathological effects due to ^{137}Cs uptake in carp (*Cyprinus carpio L.*) and eel (*Anguilla Anguilla* (*L*)). Fresenius Environm. Bull. **19**(4a), 735–738 (2010)
281. H. Florou, V. Tsytsugina, G.G. Polikarpov, F.K. Vosniakos, N. Evangeliou, A comparative study on genotoxic effects in oligochaeta from areas with elevated natural and artificial radioactivity. J. Environ. Protect. Ecol. **10**(3), 807–814 (2009)
282. F.K. Vosniakos, Radiological impact assessment of natural radioactivity concentration along the Pieria coastline in Greece. Fresenius Environ. Bull. **13**(2), 164–166 (2004)
283. H. Florou, ^{137}Cs in abiotic component and biota from Aegean and Ionian Sea—Greece. Chem. Ecol. **12**(4), 253–258 (1996)
284. A.C. Chamberlain, Interception and retention of radioactive aerosols by vegetation. Atmos. Environ. **4**, 57–78 (1970)
285. FAO, Organization of Surveys for Radionuclides in Food and Agriculture, FAO Atomic Energy Series No. 4, Rome (1962)
286. O. Paakola, Contamination of foods in Finland due to the Chernobyl accident, in *Proceedings of European Meeting of Toxicology Forum,* Geneva (1986), p. 4
287. D. Charles, M. Jones, J.R. Cooper, Radiological impact on EC Member States of routine discharges into North European waters—Report of Working Group IV of CEC Project MARINA, NRPB Report, NRPB-M172 (Chilton, Didcot, 1990)
288. H. Florou, Natural and artificial radioactivity in Marine Ecosystems, Ph.D. Thesis, University of Athens (1990)
289. G. Trabidou, H. Florou, A. Angelopoulos, L. Sakelliou, Environmental study of the radioactivity of the spas in the island of Ikaria. Rad. Prot. Dos. **63**, 63–67 (1996)
290. V.N. Egorov, G.G. Polikarpov, N.A. Stokozov, L.G. Kulebakina, G.A. Lazorenko, Distribution of artificial radionuclides in water, bottom sediments and hydrobionts of the Black Sea Basin following the Chernobyl NPP accident and assessment of ^{137}Cs input to the seas of Mediterranean Basin through Bosporus, in *Proceedings of Seminar on the Radiological Exposure of the Population of the E.C. in the Mediterranean Sea*, Marina-Med., ENEA, Rome 17–19 May 1994

Index

F. K. Vosniakos, *Radioactivity Transfer in Environment and Food*,
Environmental Science and Engineering, DOI: 10.1007/978-3-642-28741-1,